Data-Intensive Research

This book series provides a comprehensive and up-to-date collection of research and experimental works, summarizing state-of-the-art developments in the fields of data science and engineering. The trends, technologies and state-of-the art research related to data collection, storage, representation, visualization, processing, interpretation, analysis, and management related concepts, taxonomy, techniques, designs, approaches, systems, algorithms, tools, engines, applications, best practices, bottlenecks, perspectives, policies, properties, practicalities, quality control, usage, validation, workflows, assessment, evaluation, metrics, and many more are to be covered.

The series will publish monographs, edited volumes, textbooks and proceedings of important conferences, symposia and meetings in the field of autonomic and data-driven computing.

Sanjay Chakraborty · Lopamudra Dey

Computing for Data Analysis: Theory and Practices

Springer

Sanjay Chakraborty
Department of Computer Science
and Engineering
Techno International New Town
Kolkata, West Bengal, India

Lopamudra Dey
Department of Computer Science
and Engineering
Heritage Institute of Technology
Kolkata, West Bengal, India

ISSN 2731-555X ISSN 2731-5568 (electronic)
Data-Intensive Research
ISBN 978-981-19-8003-9 ISBN 978-981-19-8004-6 (eBook)
https://doi.org/10.1007/978-981-19-8004-6

This Springer imprint is published by the registered company Springer Nature Singapore Pte Ltd.
The registered company address is: 152 Beach Road, #21-01/04 Gateway East, Singapore 189721, Singapore

To our Parents, Sister and my Son Arohan for their love and inspiration.

—Dr. Sanjay Chakraborty
—Dr. Lopamudra Dey

Preface

Data analytics is significant since it aids in the performance optimization of enterprises. By finding more cost-effective ways to do business and retaining a lot of data, firms can help cut expenses by incorporating it into their business strategy. Analyzing data collections to identify trends and make judgments about the information they contain is known as Data Analytics (DA). Data analytics is increasingly carried out with the use of specialist hardware and software.

This book has covered various cutting-edge computing technologies and their applications over data. We have discussed in-depth knowledge of big data and cloud computing, the Internet of Things, augmented and virtual reality, quantum computing, cognitive computing, and computational biology with respect to different kinds of data analysis and their applications. In this book, we have described some interesting models in the cloud, IoT, AR/VR systems, quantum, cognitive, and computational biology domains that provide some useful impact on intelligent data (bulk time series data, emotional, image data, etc.) analysis. We have also explained how these computing technologies-based data analysis approaches are used for various real-life applications. We believe this book will be benefited the readers who are interested to work in these areas in the future.

Chapter 1 gives an overall introduction to the basics of big data analytics, cloud data analytics, quantum, and IoT-based data analytics, biological data analytics, and so on. It described the impact of data analysis on all these frameworks briefly.

Then under Part I, Chap. 2 discusses the roles and different techniques of big data analysis on a cloud platform. It first discusses the various types of big data analysis and how that data analysis can be performed in Hadoop architecture through a cloud framework. To find patterns in data and derive fresh insights, cloud analytics entails the combination of scalable cloud computing with robust analytical tools. Data analysis is being used by corporations to gain a competitive edge, enhance scientific research, and improve people's lives in a variety of ways. Therefore, it describes the basics of the cloud, its different models, and architectures and also explains how they help to do effective data analysis.

Chapters 3 and 4 focus on the discussion of edge computing with the notions of the Internet of Things (IoT) and augmented/virtual (AR/VR) reality. Chapter 3 introduces the basic concepts of IoT along with the related technologies, protocols, and architecture. Then, it describes the impact of IoT on various industrial applications and big data analysis on cloud framework. Chapter 4 discusses the types of augmented reality with some specific system architectures. It also explains the different hardware and software components of AR/VR systems. It also presents the different real-life applications of data analysis and future research directions in this area.

Chapter 5 under Part II takes a more in-depth look at data analysis in the Biocomputing domain. In this domain, we discuss the basic concepts of computational biology and its various data types. Besides that, it describes the different data analysis processes on DNA/RNA sequences, microarray data sequences, and protein sequences.

Chapter 6 discusses the data analysis through cognitive computing. In this chapter, we describe the basics of brain–computer interfacing techniques for feature extraction and its various components. It also presents the methodology for the classification of emotional data through the analysis of EEG signals collected from the human brain. It has huge applications for those people who are getting distressed due to work pressure or other issues in their day to day life.

In Part III, Chaps. 7 and 8 deal with the concepts of quantum computing that help to perform various machine learning and image processing operations on a set of real-life data and image matrices. Chapter 7 discusses the basics of quantum machine learning concepts and how they can be utilized to solve some complex clustering and classification problems more efficiently compared to classical computing. Similarly, two important and complex image processing operations (denoising and edge detection) that can be solved more efficiently and faster way in the quantum framework are discussed in Chap. 8.

Finally, Chap. 9 under Part IV summarizes the concepts presented in this book and discusses applications and trends in data analysis. Social impacts of data analysis, such as privacy and data security issues, are discussed, in addition to challenging research issues.

This book has several strong features that set it apart from other texts on computing for data analysis. It presents very broad yet in-depth coverage of the spectrum of data analysis over various popular computing domains, especially regarding several recent research topics on data computing.

Dr. Sanjay Chakraborty
Associate Professor
Department of Computer Science and Engineering
Techno International New Town
Kolkata, India

Dr. Lopamudra Dey
Assistant Professor
Department of Computer Science and Engineering
Heritage Institute of Technology
Kolkata, India

Acknowledgements

We express our great pleasure, sincere thanks, and gratitude to the people who significantly helped, contributed, and supported the completion of this book. We are sincerely thankful to Dr. Radha Tamal Goswami, Professor and Director, Techno International Newtown, Kolkata, India, for his encouragement, support, guidance, advice, and suggestions to complete this book. Our sincere thanks to Dr. Amlan Chakrabarti, Professor and Head, AKCSIT, University of Calcutta, India, and Dr. Anirban Mukhopadhyay, Professor, Department of Computer Science and Engineering, University of Kalyani, Kalyani, India, for their continuous support, advice, and cordial guidance from the beginning to the completion of this book.

We would also like to express our honest appreciation to our colleagues at the Techno International Newtown, India, and Heritage Institute of Technology, Kolkata, for their guidance and support.

We are also very thankful to the reviewers for reviewing the book chapters. This book would not have been possible without their continuous support and commitment toward completing the review on time.

To complete this book, the entire staff at Springer extended their kind cooperation, timely response, expert comments, and guidance, and we are very thankful to them.

Finally, we sincerely express our special and heartfelt respect, gratitude, and gratefulness to our family members and parents for their endless support and blessings.

Kolkata, India

Dr. Sanjay Chakraborty
Dr. Lopamudra Dey

Contents

About the Authors

Dr. Sanjay Chakraborty is currently an Associate Professor of the Department of Computer Science and Engineering, Techno International New Town, Kolkata, India. He did his B.Tech. from West Bengal University of Technology, India on Information Technology in the year 2009. He completed his Master of Technology (M.Tech.) from National Institute of Technology, Raipur, India in the year of 2011. He completed his Ph.D. at AKCSIT University of Calcutta in 2022. Dr. Chakraborty is the recipient of the University Silver Medal from NIT Raipur in 2011 for ranking first class second in M.Tech. He has 11 years of teaching and research experience. He has published over 55 research papers in various international journals, conferences and book-chapters. He has authored of two books published by Lap Lambert, Germany and Springer EAI series respectively. Dr. Chakraborty attended many national and international conferences in India and abroad. His research interests include Data Mining and Machine Learning and Quantum Computing. He is a professional member of IAENG and UACEE. Dr. Chakraborty is an active member of the board of reviewers in various International Journals, Transactions and Conferences. He is the recipient of "INNOVATION AWARD" for outstanding achievement in the field of Innovation by Techno India Institution's Innovation Council 2019. He is also the recipient of "IEEE Young Professional Best Paper Award" in 2017. He has also achieved the top five best paper recognition by Ain Shams Engineering Journal, Elsevier and most cited author award from Biomedical Journal, Elsevier in 2021.

Dr. Lopamudra Dey completed B.Tech. from West Bengal University of Technology, Kolkata, India in Computer Science and Engineering in 2009. She received a Bronze medal in her Bachelor degree. In 2011, she completed M.Tech. from University of Kalyani West Bengal India. She obtained her Ph.D. in Computer Science from Kalyani University in 2021. She is also working as an Assistant Professor in the Department of Computer Science and Engineering in Heritage Institute of Technology, Kolkata, India. Her areas of interests include Bioinformatics, Data Mining, and Network Security. She has published more than 15 research articles in journals, conferences and books.

Chapter 1
Introduction

1.1 Data and Analysis

Data, which is shorthand for "information", has always been gathered, reviewed, and/or analyzed as part of the running of the Head Start program. For children to enroll in the program, numerous pieces of information are needed. Information from screenings and any subsequent services are included in the delivery of health and dental services. The gathering and use of a significant amount of information are required in every aspect of a Head Start program, including content and management [1]. No matter if they identify as "data analysts" or not, everyone in today's world must cope with mountains of data. However, those that have a toolbox of data analysis abilities have a huge advantage over everyone else because they know what to do with all that information. They are skilled at turning data into knowledge that motivates practical action. They are skilled at deconstructing and organizing complicated issues and datasets to get at the root of issues in their industry.

1.1.1 Types of Data

The relative benefits of quantitative and qualitative data have been the subject of a protracted argument in the research community. Key factors in this discussion include the researchers' educational backgrounds, which are exacerbated by individual differences and people's preferences for relating to things in words or figures. In actuality, Head Start does not really care about this argument. We need to gather both kinds of data if we want to have a high-quality program.

- **Qualitative Data** is information that is conversational or narrative in nature. Focus groups, interviews, open-ended questions on questionnaires, and other less organized methods are used to gather these kinds of data. Thinking of qualitative data as words is a straightforward method to examine it.

S. Chakraborty and L. Dey, *Computing for Data Analysis: Theory and Practices*,
Data-Intensive Research, https://doi.org/10.1007/978-981-19-8004-6_1

- Data that is expressed numerically and can have either large or tiny numeric values is referred to as **Qualitative Data**. A certain category or label may be associated with a number of values.

1.1.2 Analysis of Data

The study of unstructured data is to uncover patterns and relevant knowledge. Additionally, this procedure may involve gathering, organizing, preprocessing, transforming, modeling, and interpreting the data. Knowledge in the field of analytics comes from several resources. The concept of extrapolating information originates in the long-established field of inductive learning, a subfield of statistics. With the development of personal computers, computational resources are being used increasingly frequently to address issues related to inductive learning. The ability to compute has been used to create novel techniques. New issues have also arisen that call for a solid understanding of computer sciences. For instance, computational statisticians now explore ways to carry out a specific task more efficiently from a computational standpoint.

Several scientists have also fantasized about being able to simulate human behavior on machines. They came from the artificial intelligence field. In addition to statistics, they also employed computers to simulate biological and human behavior, which was a major source of inspiration for their study. For instance, artificial neural networks have been investigated since the 1940s to mimic the human brain, and ant colony optimization algorithms were developed in the 1990s to mimic the behavior of ants. According to Arthur Samuel in 1959 [2], the term machine learning (ML) first originated as the "area of study of computer algorithms that convert data into intelligent tasks". A new phrase with a marginally different connotation first surfaced in the 1990s: data mining (DM). Business intelligence tools first became available in the 1990s as a result of more affordable and large-capacity data centers [2].

Companies begin to gather an increasing amount of data with the intention of either resolving or improving business operations, such as by identifying credit card fraud, enhancing client relationships using relational marketing strategies that are more effective. The main issue was whether it was possible to fetch the data to draw out the knowledge required for a certain purpose.

1.2 Big Data and Data Analytics

The phrase "big data" initially originated in the early twentieth century. A big data, the "three Vs" initially served as the definition of data processing technology. Since then, other Vs have been suggested. We may create a taxonomy of big data using the first three Vs: volume, variety, and velocity. There is a volume issue with data repositories for massive amounts of data as a method of storing big data. How to combine data from several sources is a topic of variety. Velocity refers to the capacity to handle data arriving quickly and in streams called data streams. Learning from streaming data outside of big data's velocity is another aspect of analytics. A new term has evolved and is occasionally used as data science. Large datasets require the development of new techniques and tools for data storage, computing, and distribution because they cannot be handled by the data processing technologies that are now available [3]. Big data, however, can be described in many ways than just data amount. The term "big" can be used to describe a variety of factors, including the quantity of data sources, the significance of the data, the demand for new processing methods, the speed at which data is received, the combination of various datasets to enable real-time analysis, and the accessibility of the data, which today is available to any business, non-profit organization, or individual. Big data is therefore more focused on technology. It offers a computer platform for various data processing operations in addition to analytics.

Processing financial transactions, processing online data, and processing georeferenced data are some of these responsibilities. Data science focuses on the development of models that can recognize patterns in large amounts of complex data and the application of these models to practical issues. Data science uses the right technology to extract meaningful and practical knowledge from data. It is closely related to data mining and analytics. By offering a framework for knowledge extraction that incorporates statistics and visualization, data science goes beyond data mining.

As a result, although data administration and collection are supported by big data, new knowledge is discovered through data science through the application of procedures to these data. All of these techniques for drawing knowledge from data are included by the concept of data analytics that we utilize [4, 5].

1.2.1 Big Data Architecture and Data Analysis

New computer technologies are required when data grow in bulk, velocity, and variety. These emerging technologies, which comprise h/w and s/w, must be highly flexible as more data is processed. Scalability is the name for this quality. Distributing the data processing jobs among a number of computers, which may then be grouped together to form computer clusters, is one technique to achieve scalability. The reader should not conflate computer clusters with clusters created by analytics techniques called clustering, which partition a dataset to locate groupings within it. Even though a distributed system can be created by grouping numerous computers into a cluster, conventional distributed system software typically struggles to handle massive data. The effective division of data among the various computing and storage units is one of the restrictions. New software tools and approaches have been created to handle these requirements. MapReduce was one of the first methods created for huge data processing employing clusters. The two steps in the MapReduce programming model are map and reduce. Hadoop is the name of the most well-known MapReduce implementation. MapReduce separates the dataset into pieces, or "chunks", and saves the block of the dataset required by each cluster computer [4]. The average salary of a billion people might be calculated using a cluster of thousand computers, each of which has a computing unit and storage capacity. The population can be broken down into 1000 subgroups, or pieces, comprising data from one million individuals each. One of the computers can process each chunk on its own. One may average the output of each of these computers, which represents the average wage of one million individuals, to obtain the final average salary. The following conditions must be met by a distributed system in order to effectively tackle a large data problem:

- Ensure that the entire task is completed and that no data is lost. Another computer in the cluster must take up the responsibilities assigned to the failed computer or computers, as well as the affected data chunk.
- Redundancy is the practice of performing the identical task and associated data piece on many cluster computers. As a result, the redundant computer continues to perform the work even if one or more computers fail.
- Faulty computers can rejoin the cluster once they have been repaired.
- As the processing demand varies, it is simple to withdraw computers from the cluster or add more ones.

A solution that complies with these requirements must conceal from the data analyst the mechanics of how the program functions, such as how the jobs and data blocks are allocated among the cluster computers [6]. Chapter 2 describes how big data analysis can be performed on a distributed cluster environment in details.

1.3 Cloud Computing and Data Analysis

To find patterns in data and derive fresh insights, cloud analytics entails the combination of scalable cloud computing with robust analytical tools. Data analysis is being used by corporations to gain a competitive edge, enhance scientific research, and improve people's lives in a variety of ways. Data analysis is being used by corporations to gain a competitive edge, enhance scientific research, and improve people's lives in a variety of ways. Consequently, as the amount and value of data continue to rise, data analytics has grown in importance as a tool. Artificial intelligence (AI), machine learning (ML), and deep learning are frequently linked to cloud analytics (DL). Additionally, it is frequently utilized in commercial applications, including corporate intelligence, security, Internet of Things (IoT), genomics research, and work in the oil and gas industry. In truth, data analytics may boost organizational performance and create new value in every sector. A subset of cloud analytics called cloud infrastructure analytics is concerned with the analysis of data related to IT infrastructure, whether it is on-premises or on the cloud. Identification of input–output patterns, performance evaluation of applications, detection of policy compliance, and support for capacity management and infrastructure resilience are the objectives [7, 8].

Data analytics, or the process of analyzing and drawing conclusions from massive datasets, has become easier because to the development of analytics programs like Apache Hadoop. Analytics workloads and technologies that were migrated to the cloud are now referred to as cloud analytics. The capability, accessibility, and ease of executing complicated data analysis on very big datasets have all grown significantly thanks to cloud analytics. For a number of reasons, cloud analytics is particularly intriguing:

- The amount of data being gathered globally is increasing at startling rates, and a large portion of it is being created and gathered at IoT endpoints or in the cloud.
- Because cloud services are supplied as automated services and do not involve the installation and upkeep of physical hardware, they are significantly simpler to deploy.
- A user can activate and deactivate services as necessary thanks to the cloud business model. With this consumption-based pricing model, clients only pay for the services they actually use, eliminating the need to purchase and manage expensive hardware and saving money on data center space.
- Users can use the cloud to deploy the ideal number of IT resources based on the current issue. Users may quickly apply computing and storage and grow them as needed thanks to dynamic resource sizing. Users are relieved of the need to purchase a fixed capacity of physical IT equipment for each project involving data analysis.
- For users that want to use the cloud to test a new analytics project as a POC before making investments on-premises, using a hybrid analytics solution is effective [9].

Organizations are empowered by cloud analytics to:

- Analyze genomic data to learn more about hereditary disorders and how to develop treatments.
- To enhance customer happiness and customer service, look for patterns in voice, photographs, and videos.
- To increase product availability and delivery, research purchasing patterns.
- Determine disease reporting patterns to increase the accessibility of medications and immunizations.
- Hybrid cloud infrastructures should be analyzed to reduce IT spending and enhance application performance.

There are some best uses of cloud data analytics given below,

A. **Social Media**
Compounding and deciphering social media activity is a common application for cloud data analytics. Processing activity across numerous social networking sites was challenging until cloud drives became widely used, especially if the data was stored on different servers. Cloud drives enable simultaneous social media site data analysis, enabling speedy results quantification and attention-based resource allocation.

B. **Tracking of Products**
It should come as no surprise that Amazon.com, long regarded as one of the kings of efficiency and foresight, employs data analytics on cloud storage to follow things across their chain of warehouses and distribute items wherever is necessary, regardless of the items' proximity to customers. With the help of their Redshift project, Amazon is a pioneer in big data analysis services in addition to using cloud drives and remote analysis. Redshift serves as an information warehouse and provides smaller organizations with many of the same analysis tools and storage capacities as Amazon. This saves smaller companies from having to invest in expensive hardware.

C. **Tracking Preference**
For the past 10 years or so, Netflix has drawn a lot of attention because to its DVD delivery service and the movie library it hosts online. One of their website's highlights is its movie suggestions, which keep note of the films users view and suggest similar ones they might like, serving as a service to customers and promoting the use of their product. All user information is remotely kept on cloud disks, so users' preferences do not alter from computer to computer. Netflix was able to produce a television program that statistically appealed to a sizable section of its audience based on their proven taste since they were able to keep all of their users' preferences and tastes in movies and television.

D. **Records Keeping Strategy**
Data may be recorded and processed simultaneously using cloud analytics, regardless of how far away local servers are. Businesses can monitor the sales of a product across all of their locations or franchisees in the USA and modify their production and shipments as necessary. They can manage inventories remotely using information that is automatically posted to cloud drives instead of waiting for inventory reports from nearby stores if a product is not selling well. Businesses can operate more effectively and have a better understanding of their customers' behavior thanks to the data stored in the cloud [10].

Chapter 2 describes how data analysis can be performed on a cloud computing environment in details.

1.4 Internet of Things (IoT) and Data Analysis

IoT analytics is a data analysis tool that evaluates the vast amount of data gathered from IoT devices. IoT analytics analyzes enormous amounts of data and generates informative data from it. IoT analytics and Industrial IoT are frequently discussed together (IIoT). Numerous sensors are used in manufacturing infrastructure, weather stations, smart meters, delivery vans, and other types of machinery to gather data. Data center management and applications for the retail and healthcare industries can both benefit from IoT analytics. IoT data, however, resembles big data. The main distinction between the two is not simply the amount of data, but also the variety of sources from which it was gathered. All of this information must be transformed into a single, understandable data stream. Data integration becomes quite challenging when there are so many different types of information sources. This is where IoT analytics may help, even though it might be challenging to build and deploy [11].

There is an unending flow of data in large amounts from a variety of devices. Without the use of hardware or infrastructure, IoT analytics assists in the analysis of this data across all linked devices. Computing power and data storage scale up or down in accordance with changes in your organization's needs, ensuring that your IoT analysis has the necessary capability [12].

(a) Collecting data from many sources, in a variety of formats, and at various frequency is the initial stage.
(b) Then, this data is processed using a variety of outside sources.
(c) After that, the data is kept in a time series for analysis.
(d) The analysis can be carried out in a variety of methods, including using machine learning analysis approaches, ordinary SQL queries, or specialized analysis tools. Numerous predictions can be made using the findings.
(e) Organizations can create a variety of systems and applications to streamline business procedures using the information they have acquired.

There are wide range of IoT devices that capture data and help to analyzed them. Some of them are wearable devices, such as smart watch, smart glasses, smart cars. There are a list of benefits that can be achieved during data analysis through IoT [13, 14].

- Greater control and visibility, which speed up decision-making.
- Growth into new markets and adaptable scaling of business requirements.
- Automation reduces operating expenses, and improved resource use.
- New revenue streams as a result of operational issues being resolved.
- Quicker answers from precisely identifying the issues.
- Earlier problem resolution and recurrence avoidance.
- Improved client experience based on research of past purchases.
- More efficient and pertinent product development.

Chapter 3 describes how data analysis can be performed through IoT devices in details. It describes the various data collection strategies through IoT devices and their architecture and protocols for data communication. This Chap. 3 also discussed the relation between IoT and cloud services for the purpose of big data analysis.

1.5 AR/VR and Data Analysis

Bar graphs and pie charts, which were once the standard tools for data visualization, are simply unable to capture the intricacy of the data that we now gather. More than simply data scientists are required to extract insights in order to fully utilize the enormous amount of data we acquire every day. One way that artificial intelligence and augmented/virtual realities might truly alter data analytics in an organization is by making data simpler to interpret even for those who do not have a background in data. Users can better understand and spot trends in data by using visualization. Users can more easily gain insights by interacting with the data with the aid of AR and VR. In conventional 2D data visualizations, it is frequently impossible to detect critical information, such as data clusters at the intersection of several dimensions. Users of AR and VR can engage with the data since it can literally surround them on all sides and be in front of, behind, above, and to either side of them. Collaboration between teams that are spread across different places can be facilitated by VR and AR [15].

Multivariate datasets are common today. Without VR and AR, it is currently impossible for humans to effectively assess the complexity of data; therefore, they must manually put up 2D representations, reports, graphs, and the like in order to try to guide decision-making. Users using VR/AR can view everything at once, giving them a comprehensive perspective on the data that is not possible with conventional methods of data presentation. The use of a human's innate ability to consider and interpret data in various dimensions is another advantage of using VR and AR for data visualization. Data properties can be communicated in a variety of positions due to the user's immersion in the data representation. Data analytics is now accessible to a wider user base than only data scientists thanks to AR/VR technology. It can make it possible for more people to be involved in keeping an eye on neural networks and machine learning models to make sure that the decisions the machines make continue to be morally righteous, just, and logical. Data analysis may also be more enjoyable thanks to AR and VR. Humans need to embrace data analysis since it is how businesses stay competitive to make decisions based on data. By "walking into the data", data analysis becomes an immersive experience that can even be enjoyable rather than a task of poring over spreadsheets and reports [16].

Analyzing data using AR and VR makes it easier to understand the data completely. In addition, we now have the ability to display complicated data structures in ways that are easier to comprehend than previously. In AR–VR systems, massive data visualization is preferred for three reasons.

- We can reduce the complexity of the data by visualizing its structures in VR and AR.
- It is a novel media with enormous promise for data visualization. It provides more organic connections, more room, multidimensionality, fewer annihilations, and many other things.
- Not just data scientists but also other people can get data analytics thanks to AR and VR technologies.

The next major use of VR technology will be combining it with big data to address the problem caused by the limitation of human perception. If the enormous amount of data generated by user interaction can be filtered into useable information, it is an incredibly great asset. In the highly competitive environment of internet enterprises, sorting this data is crucial to making wise judgments. When it comes to processing massive datasets, traditional visual representations like pie charts and diagrams in two dimensions are not cutting it. VR thus offers an optional way to review material by exploiting its immersive capabilities to handle complicated problems. Data visualization is a concept that comprises creating an immersive experience in which the information models surround you. It makes use of intelligent mapping, intelligent routines, machine learning, and natural language processing to identify important patterns and display them in the virtual world, which users may subsequently customize. The main justification and purpose for combining VR and big data is to increase the thoroughness of the enormous volume of analytical data. One business in particular has created a platform that enables users to study up to ten data pieces by fusing artificial intelligence, virtual reality, and big data [17].

In this book, Chap. 4 discusses the various types of AR–VR systems and their organization. Besides that, it also shows how the different tools and technologies of AR–VR systems help to do an effective and efficient data analysis.

1.6 Biological Computing and Data Analysis

Data analytics is the science of analyzing unprocessed data in order to make inferences about it. Any type of data can be used using these strategies to learn things that can be used to make things better. With the use of data analytics techniques, trends and indicators that could otherwise get buried in a sea of data can be found. The overall efficiency of any model can be improved by optimizing the dataset features.

1.6.1 Steps in Data Analysis

Grouping, acquiring, cleaning, translating, and analyzing raw data into useful, pertinent information that can help businesses make informed decisions are the process of data analysis. It can be explained with the following steps:

1. The first step is to understand that the data requirement and how to make groups of data.
2. The second stage of data analytics is the data collection procedure. Computers, online resources, cameras, environmental sources, and people can all be used for this, among other methods.
3. After the data collection, data is organized using software.
4. At fourth step, data cleaning is done to eliminate duplicates, missing values, and errors.

1.6.1.1 Biological Data

Biological data refers to the information that gathered from the biological organism. There exist many different types of biological data. For example, gene sequence, protein structure, mutation, gene expression, amino acids, linkages, pathways, etc. All of these data formats are extremely complicated, and traditional database management systems (DBMS) do not adequately address the need for complex data structure as compared to most other applications. Bioinformaticists collect these biological data, mainly DNA, RNA, and protein data from computational and laboratory experiments and also published literatures, and store them in databases.

Biological data has a number of unique properties that make it difficult to manage. It has a lot of variability and a wide range. Moreover, different biologists represent the same data differently. For example, a protein name and its ID is different in different databases. The same protein SUMO1 has several aliases like DAP1, GMP1, OFC10, PIC1, SENP2, SMT3, SMT3C, SMT3H3, UBL1. It has ID 7341 in NCBI database and ID P63165 in UniProt Database. Furthermore, the schemas of biological databases are rapidly changing. There should be support for schema evolution and data object migration so that information can move more freely between database generations or releases. As most biologists have very little knowledge about the internal schema design, the interface to the biological database/resource should display information to the user in a manner appropriate for the problem being addressed and that reflects the underlying data structures. Access to past versions of existing data is frequently required by biological data users. Therefore, while updating the existing database, handling of the old data needs to be carefully managed. Finally, users of biological database need only read access and do not require write access. Write access is restricted to authorized users known as curators. Although only a small number of users require write access, the users generate a wide range of read access patterns in the databases.

1.6.1.2 Types of Biological Data and Databases

Biological databases can essentially be divided into the following groups based on the sorts of data stored in them: (1) DNA, (2) RNA, (3) protein, (4) expression, (5) pathway, (6) gene ontology. There are different biological databases that contain different biological data. For example, nucleic acid databases contain DNA information, genomic databases contain gene-level information, protein information is available at protein databases, and protein families, domains, and functional sites contain classification of proteins and domain-related data. These databases serve as repositories of biological data to researchers. Each entry in the database contains information about the nucleotide sequence, protein sequence, 3D structure, etc. A defined algorithm is required to analyze the contents of a database.

1.6.1.3 Data Analysis on Biological Data

Over the last decade, biological data is growing rapidly. Human genomes can now be sequenced 50,000 times quicker than they could in 2000. As biological data volumes increase, existing analysis techniques and environment can no longer keep up with the demand for data analysis activities to be completed quickly in the life sciences. Three key characteristics of biological datasets are enormous data volume, extraordinarily long running time, and application reliance. Each day, hundreds of TB of data are created. Such a vast volume of data presents problems for hardware support as well as computer scientists' ability to analyze data effectively and efficiently. Therefore, the development of efficient and effective biological data analytics technologies has required significant research investment [18].

High-performance computing (HPC) platforms and effective, scalable algorithms can provide efficient way to solve these problems. Large-scale data can be mined for useful insights thanks to data science. Principal component analysis, linear regression, and linear discriminant analysis were initiated by many of the inventors of modern statistics, such as Galton, Pearson, and Fisher, who were also preoccupied with the analysis of significant volumes of biological data [19]. Methods including logistic regression, clustering, random forests, and neural networks were envisioned or developed more recently by scientists that can solve biological issues. Apart from that, in order to take advantage of the variety of parallelism and scalability on computer platforms, various programming models such as OpenMP, CUDA/OpenCL, message passing (MPI), and MapReduce (Hadoop, SPARK) have been used by biological data researchers in many applications [20]. For networked computing, MPI is the most widely used programming paradigm. Researchers employ MPI to build high-performance biological data analytics tools on supercomputers. However, due to the strict requirements of scalability and fault tolerance (changing it), new programming models, such as MapReduce and Spark, are proposed for large-scale distributed computing [21].

1.7 Cognitive Computing and Data Analysis

Systems for cognitive computing are frequently employed to complete tasks that call for the analysis of enormous volumes of data. For instance, cognitive computing in computer science helps with large data analytics, seeing trends and patterns, comprehending human language, and connecting with clients. Cognitive analytics combines several cognitive technologies, such as semantics, artificial intelligence algorithms, deep learning, and machine learning, to do some jobs with intelligence akin to that of a human [22].

The development of big data has been the subject of numerous research that have gathered a variety of academic sources. When big data analytics is applied, cognitive computing can help minimize their drawbacks. In order to simulate both the human thought process and the system errors makes repeatedly, cognitive computing uses a computational model. This learning method can greatly improve how enormous amounts of data are analyzed for better decision-making. The first step toward advancement is implementing cognitive computing to evaluate huge data, so researching and comprehending this topic are crucial [23]. These systems deliver higher-quality services including emotional contact, cognitive health care, and automated driving. Cognitive computing has not received much attention prior to the big data era. However, the development of cognitive computing has now benefited from the growth of cloud-based AI [24]. While big data analytics offers ways to explore new data-related opportunities, cloud computing and the Internet of Things can provide s/w and h/w-dependent cognitive computing. Human big data thinking is one of the connections between big data analysis and cognitive computing. The primary distinction between big data analysis and cognitive computing is how data is processed in accordance with the human brain. Here, the machine must possess the same data ideas as people in order to comprehend information about the surroundings [25]. The cognitive system architecture with the notion of cloud and big data frameworks is shown in Fig. 1.1.

There is a list of features of big data and cognitive computing that are mapped to each other (Table 1.1).

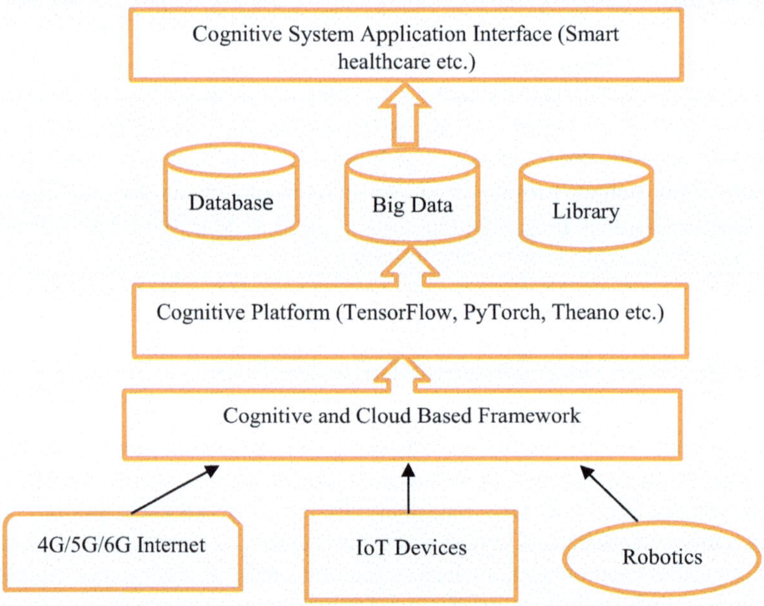

Fig. 1.1 Cognitive system architecture with cloud and big data frameworks

Table 1.1 Mapping features between big data and cognitive system

Cognitive computing features	Features
Observations	Volume
Interpretation	Variety
Evaluation	Velocity
Decision	Veracity

A cognitive computing system must be able to see a certain volume of data. To improve data analysis, a cognitive computing system can manage, purge, and normalize them. In the presence of several information sources, interpretation helps with understanding and solving difficult situations. According to variety, data may be obtained in many different ways, including through social media, IoT, GPS tracker, email services, and other channels.

A human being's innate capacity to generate knowledge includes evaluation. The cognitive computing system must evaluate massive amounts of data in a relatively short amount of time. Big data has the characteristic of velocity, wherein data creation control and processing speed are crucial. Meanwhile, the effectiveness of data analysis must be taken into account in order to produce a trustworthy and correct evaluation. Veracity is concerned with data dependability, uncertainty, and quality prediction. The term "decision feature" describes a cognitive computing system's capacity to decide in accordance with the data under analysis. The presence of evidence is one of the key factors in decision-making. Finally, the value characteristic demonstrates the futility of massive amounts of data prior to their transformation into knowledge. This function can enhance processing for knowledge development and repurpose data.

Volume, variety, velocity, veracity, and value are the five categories into which the features needed for the effective use of big data analysis are divided. The designation 5 V is used to refer to these five groups. These characteristics apply to both big data and cognitive computing. Businesses may be more productive and enterprise-ready if they can use cognitive computing to manage big data features (5 V). While conceptual modeling of structural equations has been employed, industrial big data analysis is crucial for the cognitive IoT, incorporating WSN, intellectual computer methods, and ML approaches. A cognitive computing system built on big data called the hybrid fuzzy multiobjective optimization algorithm is used to optimize social media analysis [26]. To address the *E*-projects portfolio selection problem, this approach is suggested (EPPS). Web development environments place a great deal of importance on big data decision-making in EPPS. It has been demonstrated that big data and cognitive computing are useful in the process of learning [27]. In this study, an intelligent model is utilized to investigate how big data and cognitive systems might be improved in order to redesign the labor market and have an impact on educational processes. In addition, suggestions have been made to enhance the performance of universities in order to address the issues that currently present in education. The devised remedy is predicated on a novel paradigm known as the Smart University, where knowledge expands quickly, is freely distributed, and is seen as a shared heritage of instructors and students. The key finding for a great demand for competences and expertise motivates the educational system to incorporate other disciplines into the curriculum. Here, big data use and cognitive computing systems help speed up the process of restoring key academic community components. Figure 1.2 illustrates how the advantages of big data are used to link the characteristics of cognitive computing.

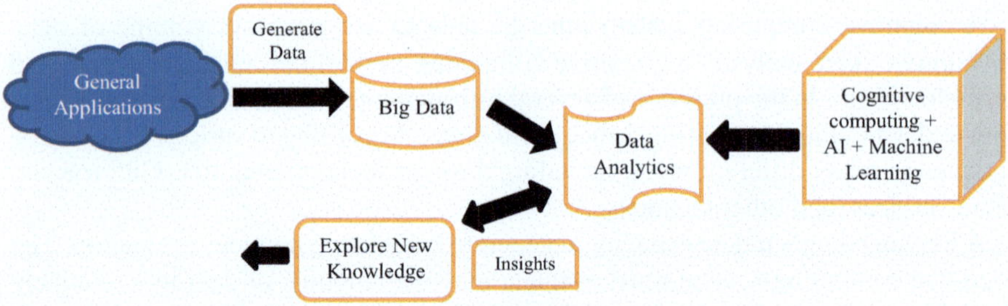

Fig. 1.2 Cognitive computing and big data-based conceptual model

Data classification through cognitive computing (brain–computer interfacing) techniques are discussed and analyzed at Chap. 6 in this book. This chapter is fully focused on the discussion of the EEG signal-based emotional data classification.

1.8 Quantum Computing and Data Analysis

Quantum computing is one of the most powerful concepts nowadays that can handle large volume of complex data collected from different scenarios efficiently and effectively. The term "big data" is a matter of concern nowadays. To handle such kind of data, we require more powerful systems, tools, and technologies. The term "big data" is frequently used interchangeably with "artificial intelligence", which leads to a misunderstanding that it refers to a problem rather than a solution. "Big data" may be a computational challenge in the field of medicine specifically if the size of a given dataset exceeds the processing capabilities of current computers. For instance, genetic data contains millions of SNPs and other biomarker data, necessitating a sizable amount of storage space and computational skill to execute studies with a semblance of efficiency. This problem is only made worse by the expanding volume of multidimensional data that is now available to study intricate phenotypes, risk factors, and outcomes. A significant gap exists in the development of cutting-edge genetic and/or molecular epidemiological research due to the limitations of traditional computing. The limitations of today's sophisticated computers are fortunately being overcome by better computing approaches like parallel processing and supercomputing. Research into quantum phenomena and optimization theory has also aided in the development of computing theory, which is now starting to come to fruition [28]. At atomic scales, quantum computing adheres to the quantum mechanical rules, which is radically different from the world as we know it. The smallest unit of information in a classical computer is a bit, a binary digit that is deterministically represented as either "0" or "1", whereas the closest equivalent unit in a quantum computer is the qubit, a 2 quantum system probabilistically represented as a coherent superposition of both "0" and "1". There are a list of quantum algorithms which are very popular and extensively use in various data analytics or predictive analytics applications. All these algorithms follow the basic quantum phenomena such as superposition, parallelism, entanglement, Grover's operation, quantum operators [29]. The most widely used quantum algorithms are listed below:

- **Supervised Quantum Learning**: The best illustration of a supervised quantum algorithm is the quantum neural network (QNN). Researchers have proposed the concept of a quantum neuron, which is built on a quantum circuit that can naturally imitate the cutoff stimulation of neurons and the feedback from various ANN configurations. Their suggested model can be utilized to build a variety of classical network configurations, including supervised, unsupervised, and reinforcement learning, while also honoring intrinsic quantum benefits, such as superposition of inputs, coherence, and entanglement. To connect machine learning and quantum computation, a decision tree classifier in the quantum realm. The paper introduces the quantum entropy impurity criterion for selecting the split node. The training data was then clustered into subclasses to enable the quantum decision tree to control quantum states by using a fidelity measure between two quantum states. In the instance of a quantum SVM, the classical data $x^{\rightarrow}$ was solely translated into quantum states using the quantum feature maps $V(\Phi(x^{\rightarrow}))$ and the kernel of the SVM was constructed from these quantum states. The quantum SVM can be trained in the same manner as a conventional SVM after the kernel matrix has been computed on the quantum computer. The quantum kernel concept is identical to the classical instance. We now use the quantum feature maps to calculate the inner product of the feature maps $(x^{\rightarrow}, ^{\rightarrow}) = |\langle \Phi(x^{\rightarrow}) | \Phi(z^{\rightarrow}) \rangle|^2$. The concept is that we might gain a quantum advantage if we select a quantum feature map that is difficult to simulate with a classical computer. Every internal node in a quantum decision tree divides the training dataset into two or more subgroups based on a particular discrete function [30]. The term "quantum decision tree" is occasionally used to describe a quantum query algorithm or quantum black box algorithm that uses quantum superposition to calculate the function $f: \{0, 1\}^n \rightarrow \{0, 1\}$. In reality, these quantum algorithms are not trees. Because they can handle nonlinearity and pooling operations, quantum convolutional neural networks (QCNN) can emulate the behavior of traditional CNN, capable of handling larger or deeper inputs and providing more sophisticated kernels. Their method is distinctive because it uses a novel quantum tomography technique that reduces system complexity by more reliably extracting the most important data.
- **Unsupervised Quantum Learning**: The two main types of unsupervised quantum machine learning techniques are dimensionality reduction and clustering algorithms. Because the database containing the vectors to be grouped requires less calls overall because to the usage of quantum algorithms, privacy enhancement is one application where quantum clustering methods can be useful. As a result, the user of the algorithm is exposed to less data from the database. Since QML algorithms can handle these problems in both vector number and dimension in logarithmic time, they outperform traditional methods exponentially in speed. Three quantum algorithms that could replace elements of classical algorithms and outperform classical algorithms in terms of speedup in clustering are quantized

and assume the existence of a black box quantum circuit that serves as a distance oracle and provides the distance between vector inputs. Their respective subroutines can be used to: (1) find the two vector dataset points that are the furthest apart from one another; (2) find the n vector dataset points that are the closest to a given point; and (3) produce neighborhood graphs of vector datasets, all in times faster than their classical counterparts. They suggest the following strategies for quantizing based on these capabilities: (1) divisive clustering, (2) K-medians clustering, and (3) unsupervised learning algorithms. Grover iterations are used by these subroutines, which are based on Grover's algorithm, to separate desirable outputs from the outcomes of computations with super positioned inputs. The visual technique dynamic quantum clustering (DQC) is effective for handling large and highly dimensional data. Its hallmark is its ability to work with large, high-dimensional datasets by exploiting differences in the density of the data (in feature space) and revealing subsets of the data. The result of a DQC analysis is a movie that demonstrates how and why sets of data points are genuinely categorized as members of simple clusters when they display correlations among all the measured variables [31]. Support vector clustering (SVC) links data points to Hilbert space states. These states can allow for the weighting of specific locations to give them more prominence, presumably as cluster center possibilities. They are represented by Gaussian wave functions. This is useful if one uses a method like SVC that can be improperly influenced by outliers. With the addition of this information, the influence of these outlier sites on computations for determining cluster centers might be weighed [31].

- **Variational Quantum Eigensolver (VQE)**: A hybrid quantum/classical approach called the Variational Quantum Eigensolver (VQE) can be used to determine the eigenvalues of a (typically enormous) matrix H. H is often the Hamiltonian of some system when this approach is applied in quantum simulations. In this hybrid algorithm, a conventional optimization loop is conducted inside of a quantum subroutine [32]. The overall circuit diagram is shown in Fig. 1.3.

  There are two essential steps in a quantum subroutine:

  – Prepare the ansatz, also known as the quantum state $|\Psi(\text{vec}(\theta))\rangle$.
  – Calculate the value of expectation $\langle\Psi(\text{vec}(\theta))|H|\Psi(\text{vec}(\theta))\rangle$.

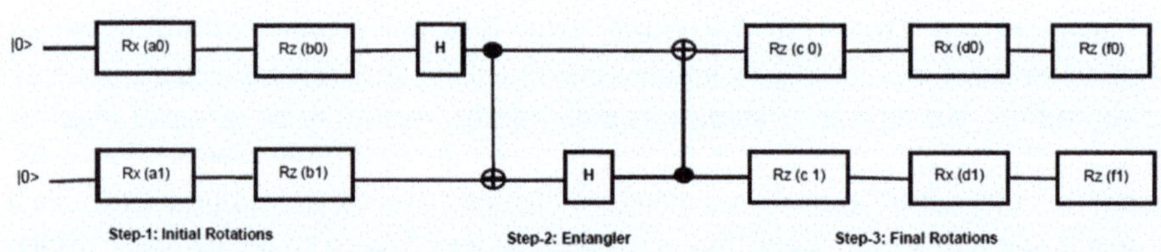

Fig. 1.3 Overall circuit diagram of VQE

This expected value will always be higher than the smallest eigenvalue of H because to the variational principle. This constraint enables us to find this eigenvalue using classical computation to execute an optimization loop:

- By adjusting the ansatz parameters vec(θ), use a traditional nonlinear optimizer to minimize the expected value.
- Until convergence, iterate.

Applications

- Solve electronic structure problems.
- In quantum chemistry to find the ground energy state of a molecule (reaction rates, binding strengths, or molecular pathways).
- Traveling salesman problem. Variational principle.
- Solve coloring puzzle (graph coloring).

- **Quantum Approximate Optimization Algorithm (QAOA)**: A variational quantum technique called the quantum approximate optimization algorithm (QAOA) is used to roughly solve discrete combinatorial optimization issues. The optimization framework of VQE is immediately extended by the QAOA implementation. However, QAOA employs its own finely calibrated ansatz, which consists of parameterized global rotations and various Hamiltonian parameterizations of the issue, in contrast to VQE, which may be configured with any number of ansatzes. The quantum approximation optimization algorithm (QAOA) is a broad method for approximating solutions to combinatorial optimization problems, especially those that may be recast as the search for an ideal bit string [33].

1.8.1 Quantum-Inspired Data Analytics

Although it is a relatively new technology, data analytics is already utilizing quantum computing. Here are a few ways that big data analysts are using quantum computing to their advantage [30, 34]:

- Quantum computing provides high-speed detection, analysis, integration, and diagnosis capabilities when working with large, dispersed datasets.
- Quantum computers can discover patterns quickly in large, unsorted datasets by concurrently observing every item in a massive database.
- Quantum computers can do incredibly complex computations in a matter of seconds as opposed to non-quantum computers, which could take hundreds of years.
- Applications of artificial intelligence employed nowadays are frequently used to manage huge data and assist in the analysis of datasets to find regularities. Despite the technology's rapid advancement, conventional computers are only capable of processing a finite amount of data. Contrarily, quantum computers are unaffected by this restriction.

Three domains of artificial intelligence benefit from the speed and power of quantum computing:

- **Natural Language Processing**: The first natural language processing operation using quantum technology was completed in 2020. Grammatical statements have been successfully converted into quantum circuits by scientists. These algorithms were able to answer questions once they were run on a quantum computer, which has significant implications for huge data.
- **Quantum Machine Learning**: Uses a quantum computer to carry out machine learning algorithms. Processing speed can be significantly increased by using this new technology, which can access more computational power than it could on a conventional computer.
- **Data Analytics for Prediction**: Using artificial intelligence, predictive analytics can be utilized to extract pertinent historical information and current data from databases. More data is processed when quantum computing is integrated with it, producing pertinent data that can then be utilized to generate predictions. However, a predictive model, which must take into consideration multiple choices, features, and variables, may find the vast amount of data accessible to be too much at times. Building more scalable predictive models with quantum computing is possible without experiencing any process sluggishness.

Environmental factors like temperature changes or vibrations can prevent most quantum computers from reaching their full potential and can place them in a condition of decoherence that renders them essentially worthless. Because of this, it might still be some time before quantum computing enters the majority of businesses or turns into a commonplace tool for data analytics. Quantum computing is still a fairly young technology in 2021. Machine learning algorithms are currently getting better thanks to developments in quantum computing. There is still a lot to be discovered about the potential of quantum computing and its implications [32].

In this book, Chaps. 7 and 8 fully deal with the different applications of quantum computing for machine learning algorithms and image processing techniques, respectively. These two chapters mainly focused on some well-known machine learning and image processing algorithms which are frequently used for different kind of general data or image matrix analysis. How quantum computing techniques help to reach an effective, efficient, and fastest data analysis is discussed in these two chapters.

1.9 Conclusion

With predictive analytics, data stream ingestion, and recommendations for critical modifications, these cutting-edge technologies are altering the data-driven enterprises. This chapter gives an overview of data analysis techniques and how the cutting-edge computing platforms enhance the efficiency, flexibility, reliability, and security of this analysis. In this book, we will concentrate on helping executives who have a lot of experience using analytics to make important business choices develop these advanced abilities.

References

1. Moreira J, Carvalho A, Horvath T (2018) A general introduction to data analytics. Wiley
2. Richmond B (2006) Introduction to data analysis handbook. Academy for Educational Development
3. Kambatla K, Kollias G, Kumar V, Grama A (2014) Trends in big data analytics. J Parallel Distrib Comput 74(7):2561–2573
4. Prabhu CSR, Chivukula AS, Mogadala A, Ghosh R, Livingston LM (2019) Big data analytics. In: Big data analytics: systems, algorithms, applications. Springer, Singapore, pp 1–23
5. Azeem M, Haleem A, Bahl S, Javaid M, Suman R, Nandan D (2021) Big data applications to take up major challenges across manufacturing industries: a brief review. Mater Today Proc
6. Shehab N, Badawy M, Arafat H (2021) Big data analytics and preprocessing. In: Machine learning and big data analytics paradigms: analysis, applications and challenges. Springer, Cham, pp 25–43
7. Ageed ZS, Zeebaree SR, Sadeeq MM, Kak SF, Yahia HS, Mahmood MR, Ibrahim IM (2021) Comprehensive survey of big data mining approaches in cloud systems. Qubahan Acad J 1(2):29–38
8. Duan L, Da Xu L (2021) Data analytics in industry 4.0: a survey. Inf Syst Front 1–17
9. Mushtaq MS, Mushtaq MY, Iqbal MW, Hussain SA (2022) Security, integrity, and privacy of cloud computing and big data. In: Security and privacy trends in cloud computing and big data. CRC Press, pp 19–51
10. Mohan PM (2021) Challenges in big data analytics and cloud computing. Int J Bus Manag Res 9(2):156–161
11. Talebkhah M, Sali A, Marjani M, Gordan M, Hashim SJ, Rokhani FZ (2021) IoT and big data applications in smart cities: recent advances, challenges, and critical issues. IEEE Access 9:55465–55484
12. Li W, Chai Y, Khan F, Jan SRU, Verma S, Menon VG, Li X (2021) A comprehensive survey on machine learning-based big data analytics for IoT-enabled smart healthcare system. Mobile Netw Appl 26(1):234–252
13. Bi Z, Jin Y, Maropoulos P, Zhang WJ, Wang L (2021) Internet of things (IoT) and big data analytics (BDA) for digital manufacturing (DM). Int J Prod Res 1–18
14. Sharma R, Sharma D (2022) New trends and applications in internet of things (IoT) and big data analytics. ISBN: 978-3-030-99329-0
15. Sharma L, Anand S, Sharma N, Routry SK (2021) Visualization of big data with augmented reality. In: 2021 5th international conference on intelligent computing and control systems (ICICCS). IEEE, pp 928–932
16. Olshannikova E, Ometov A, Koucheryavy Y et al (2015) Visualizing big data with augmented and virtual reality: challenges and research agenda. J Big Data 2:22. https://doi.org/10.1186/s40537-015-0031-2
17. Khalid ZM, Zeebaree SR (2021) Big data analysis for data visualization: a review. Int J Sci Bus 5(2):64–75
18. Venter JC (2010) Multiple personal genomes await. Nature 464(7289):676–677
19. Yin Z, Lan H, Tan G, Lu M, Vasilakos AV, Liu W (2017) Computing platforms for big biological data analytics: perspectives and challenges. Comput Struct Biotechnol J 15:403–411
20. Fienberg SE (1992) A brief history of statistics in three and one-half chapters: a review essay. Stat Sci 7:208–225
21. Langmead B, Schatz MC, Lin J, Pop M, Salzberg SL (2009) Searching for SNPs with cloud computing. Genome Biol 10(11)
22. Hurwitz JS, Kaufman M, Bowles A (2015) Cognitive computing and big data analytics. Wiley, p 288. ISBN: 978-1-118-89662-4
23. Mishra S, Tripathy HK, Mallick PK, Sangaiah AK, Chae GS (eds) (2021) Cognitive big data intelligence with a metaheuristic approach. Academic Press
24. Sechin Matoori S, Nourafza N (2021) Big data analytics and cognitive computing: a review study. J Bus Data Sci Res 1(1):23–32

25. Sreedevi AG, Harshitha TN, Sugumaran V, Shankar P (2022) Application of cognitive computing in healthcare, cybersecurity, big data and IoT: a literature review. Inf Process Manage 59(2):102888
26. Sangaiah AK, Goli A, Tirkolaee EB, Ranjbar-Bourani M, Pandey HM, Zhang W (2020) Big data-driven cognitive computing system for optimization of social media analytics. IEEE Access 8:82215–82226
27. Coccoli M, Maresca P, Stanganelli L (2017) The role of big data and cognitive computing in the learning process. J Vis Lang Comput 38:97–103
28. Mallow GM, Hornung A, Barajas JN, Rudisill SS, An HS, Samartzis D (2022) Quantum computing: the future of big data and artificial intelligence in spine. Spine Surg Relat Res 6(2):93–98
29. Shaikh TA, Ali R (2016) Quantum computing in big data analytics: a survey. In: 2016 IEEE international conference on computer and information technology (CIT). IEEE, pp 112–115
30. Chen SYC, Wei TC, Zhang C, Yu H, Yoo S (2022) Quantum convolutional neural networks for high energy physics data analysis. Phys Rev Res 4(1):013231
31. Ramezani SB, Sommers A, Manchukonda HK, Rahimi S, Amirlatifi A (2020) Machine learning algorithms in quantum computing: a survey. In: 2020 international joint conference on neural networks (IJCNN). IEEE, pp 1–8
32. Ostaszewski M, Trenkwalder LM, Masarczyk W, Scerri E, Dunjko V (2021) Reinforcement learning for optimization of variational quantum circuit architectures. Adv Neural Inf Process Syst 34:18182–18194
33. Wang H, Zhao J, Wang B, Tong L (2021) A quantum approximate optimization algorithm with metalearning for MaxCut problem and its simulation via TensorFlow quantum. Math Probl Eng
34. Pandey A, Ramesh V (2015) Quantum computing for big data analysis. Indian J Sci 14(43):98–104

Part I
Integration of Cloud, Internet of Things, Virtual Reality and Big Data Analytics

Chapter 2
Impact of Big Data and Cloud Computing on Data Analysis

2.1 Big Data Architecture with Hadoop and MapReduce

Big data deals with large volume of multidimensional data such as the data generated by Google, Yahoo, LinkedIn, eBay, etc. Big data analytics defines the techniques by which this huge amount of data can be processed and analyzed in a rapid and cost-effective manner. A traditional database management system (DBMS) fails to handle such big data. Therefore, Google develops its own MapReduce technique that can efficiently works on Google File System. Due to the BigTable system embedded into Google MapReduce framework, it becomes easy for searching from millions of data and returning the result in milliseconds. The characteristics of big data lie on three pillars velocity, variety and volume (stored in data warehouses). The benefits of using this big data concept are listed below.

- You can get more comprehensive answers thanks to big data because you have access to more data.
- More thorough responses increase data confidence, which calls for an entirely different strategy for approaching issues.
- Big data's capacity to assist businesses in product innovation and redesign is a tremendous advantage.
- Big data analytics is utilized to provide marketing insights and solve problems for advertisers.
- Businesses can detect a variety of customer-related patterns and trends thanks to the utilization of big data. By analyzing customer's purchasing behavior, a business can research the most popular products and develop items in line with this pattern.
- Big data tools can handle and analyze the customer feedback about the company through sentiment analysis, this leads to managing and increasing the growth of your business.
- Hadoop and MapReduce tools can find new data sources that assist firms in speedy data analysis and decision-making based on the knowledge.

S. Chakraborty and L. Dey, *Computing for Data Analysis: Theory and Practices*,
Data-Intensive Research, https://doi.org/10.1007/978-981-19-8004-6_2

There are three varieties of big data [1].

i. **Structured Big Data**: Structured data that has a specific format and can be processed, saved, and retrieved. It alludes to highly organized data that can be quickly and easily stored in a database and accessed from it using basic search engine methods. For example, the student table in an institute database will be structured and organized.
ii. **Unstructured Big Data**: Data that has no form or organization at all is referred to as unstructured data. Processing and analyzing unstructured data become extremely challenging and time-consuming as a result. An example of unstructured data is email.
iii. **Semi-structured Big Data**: Data that contains both of the aforementioned formats, i.e., structured and unstructured data, is referred to as semi-structured data. To be exact, it refers to data that has essential information or tags that separate different data items even though it has not been categorized under a certain repository (database). The activities performed of big data can be classified into three categories, **store** the big data in a distributed environment, cleaning, modifying, transforming, running of algorithms under a **process** is a complex job, and lastly, **access** (search, retrieval) the data. However, there are a certain number of analytics on big data,

i. **Descriptive Analytics**

It is focused fully on historical data. In this case, data warehousing plays a vital role to store the last 10–15 years of historical data. Data aggregation and data mining tools are used in this kind of analytics to discover patterns from the historical data.

ii. **Predictive Analytics**

It is a collection of statistical methods that deals with the machine learning algorithms find trends in data and forecast the future behavior and actions. Predictive analytics software is no longer just for statisticians; it is now more readily available and less expensive for a variety of sectors and industries, including the field of learning and development.

iii. **Prescriptive Analytics**

Prescriptive analytics is a statistical technique for formulating advice and taking judgments based on the results of computations from algorithmic models. Without knowing what to look for or what problem has to be fixed, recommendations cannot be generated. Prescriptive analytics starts with a problem in this manner. For example, using predictive analysis, a training manager learns that the majority of students who lack a specific talent would not finish the recently launched course. What is possible to do? Prescriptive analytics can now help with the situation and help choose options for action. Perhaps an algorithm can identify students who need the new course but lack a specific talent and automatically suggest that they use a different training resource to pick up the deficient skill. However, the correctness of a given conclusion or suggestion depends on how well the computational models and the data were developed. When implemented in the training department of another organization, what might make sense for one company's training requirements might not make sense for another. It is generally advised that models be customized for each particular circumstance and requirement.

2.1.1 Hadoop Architecture

Hadoop is a Java-based open-source software framework (MapReduce) from Apache that efficiently handle big data especially unstructured data. It is actually running on a cluster of machines to manage and store big data in a distributed manner. Besides Google, today many companies using Hadoop with MapReduce to handle big data activities inside their organization. Figure 2.1 represents all the important components of Hadoop architecture. The Hadoop architecture mainly consists of,

- MapReduce.
- Hadoop distributed file system (HDFS).
- Yet another resource negotiator framework (YARN).
- Hadoop Common.

Hadoop ecosystem plays a vital role to meet the needs of big data processing. It includes, HDFS, HBase, Hive, Sqoop, Flume, Spark, MapReduce, Pig, Impala, Cloudera, Oozie, Hue. In this chapter, we are mainly focusing on MapReduce component [2].

A. **MapReduce**

MapReduce is nothing more than a YARN framework-based data structure. The primary function of MapReduce is to carry out parallel distributed processing in a Hadoop cluster, which is what makes Hadoop operate so quickly. Serial processing is no longer useful when working with big data. It divides into Map and Reduce in these two phases. From Fig. 2.2, it is clearly visible that big data

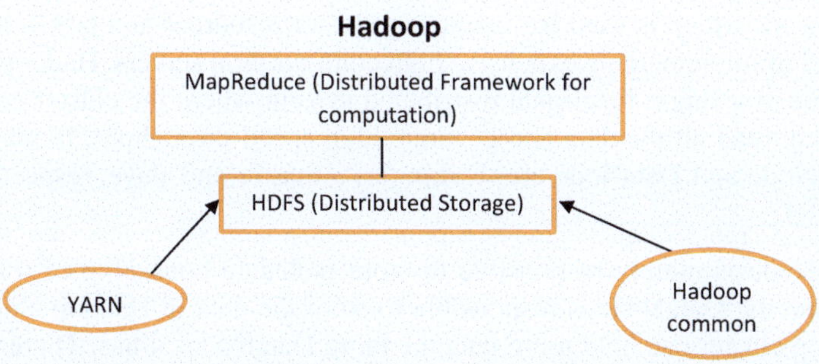

Fig. 2.1 Components in Hadoop architecture

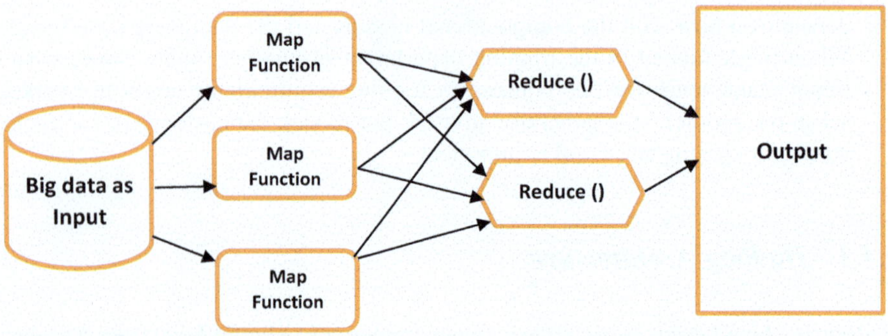

Fig. 2.2 Overall MapReduce operation phases

input is initially accepted by Map() function which divides the data into key-value pair-based tuples with the help of RecordReader module. These tuples will act as input to the Reduce() function. Reduce() merge those individual tuples into a set of tuples by its key value through a combiner module. However, some basic operations like shuffling, summation, sorting etc., can be executed on those set of tuples based on the requirements and finally send it to the output. Gathering the tuple produced by Map and performing some sort of aggregation operation on those key-value pairs relying on their key element is the primary duty or function of Reduce. In the output phase, with the aid of record writer, the key-value pairs are entered into the file with each record starting on a new line and the key and value separated by spaces [3].

B. **Hadoop Distributed File System (HDFS)**

Based on the Google File System (GFS), the Hadoop distributed file system (HDFS) offers a distributed file system that is intended to function on common hardware. It is meant to be installed on inexpensive hardware and is extremely fault-tolerant.

It supports applications with massive datasets and offers high throughput access to application data. It consists of two modules, Hadoop Common is nothing but Java libraries which are useful in other modules and Hadoop YARN is a framework which is used for managing cluster resources and task scheduling. Instead of implementing expensive high configuration servers, Hadoop helps to implement a single functional distributed system where the cluster computers parallely read all the data and generate high speed throughput. In this cluster, NameNode and DataNode are working as a master and slave, respectively [4] (Fig. 2.3).

NameNode is mainly responsible to store metadata consisting of transaction logs and the DataNode is responsible to store the data in Hadoop cluster. The Hadoop cluster can hold more data the more DataNodes it has. Therefore, it is recommended that the DataNode has a high storage capacity in order to store

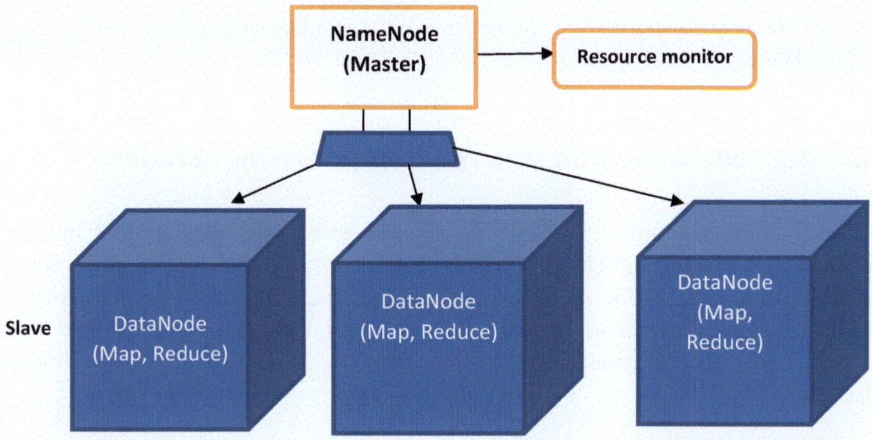

Fig. 2.3 Hadoop HDFS architecture

a lot of file blocks. HDFS stores data in terms of blocks at all times. Hadoop performs the below tasks,

- Initially, directories and files are used to organize data. Blocks of 128 and 64 M uniformly sized files make up each file (preferably 128 M).
- These files are then split up across other cluster nodes for additional processing.
- The processing is under the supervision of HDFS, which sits atop the local file system.
- Block replication is done to handle hardware failure.
- Verifying that the code was successfully executed.
- Executing the sort that comes after the map and before the reduce phases.
- Delivering the data after sorting to a certain machine.
- Generating debugging logs for every task.

C. **YARN (Yet Another Resource Negotiator)**

MapReduce runs on a framework called YARN. The two tasks that YARN carries out are resource management and job scheduling. The goal of job scheduling is to break large tasks down into smaller ones so that each job can be distributed across different slaves in a Hadoop cluster, maximizing processing. The job scheduler also keeps track of the jobs' priorities, dependencies on one another, importance levels, and other details like job timing. To manage all the resources made available for running a Hadoop cluster, Resource Manager is used.

D. **Hadoop Common**

Hadoop Common, often known as the "common utilities", is nothing more than our Java library, Java files, or the Java scripts that we require for all the other components found in a Hadoop cluster. For the cluster to function, HDFS, YARN, and MapReduce use these tools. Hadoop Common confirms that hardware failure in a Hadoop cluster is frequent, necessitating an automatic software solution by the Hadoop Framework.

2.2 Big Data Analytics: Emerging Applications in Industry

Today is an era of big data. The notable improvement of the bandwidth of Internet or the use of Internet of Things (IoT) helps the rapid growth of big data. These days, any firm is amassing multidimensional data, including infobytes from media to journal articles, tweets to YouTube videos, social networking updates, and blog conversations. There are multiple industries using big data applications on the cloud platform.

1. **Financial Institution and Banking**

   Big data is extensively used to monitor the activity of financial markets like the stock exchange nowadays. Huge data is utilized by retail traders, big banks, hedge funds in the financial markets for trade analytics such as high-frequency trading, sentiment analysis, predictive analytics. For risk analytics, such as antimoney laundering, demand enterprise risk management, "KYC", and detection of fraud extensively relies on big data.
2. **Healthcare Industry**

   Some hospitals are trying to collect feedback about the doctors, infrastructure, facilities, and so on from the patients and their guardians through mobile applications. Based on that, they can improve their services to the patients. Some medical institutes have combined free public health data and Google Maps to provide visual data that enables quicker diagnosis and effective analysis of healthcare information used in tracing the development of chronic disease.
3. **Education Sector**

   Nowadays, big data has been extensively used in education sectors. Some institutes monitor the overall progress of the students over time through the use of big data-based learning and management system. It is also used to measure the teaching quality, performance, and effectiveness of the teachers or trainers to ensure gradual growth for the students.
4. **Media, Entertainment and Social Networking**

   All kinds of social media and entertainment industries (YouTube, Facebook, Netflix, Amazon Prime, etc.) are using big data techniques to generate content for various target audiences, provide on-demand content, and also monitor the quality of the content. Real-time sentiment analysis during a football match or cricket match can be possible through big data. Nowadays, a lot of work is going on "recommendation systems" where big data applications play a significant role.

5. **Retail and Online Shopping**

   Big data from social media is used for product promotion, customer interest, and customer retention. Big data technology also helps these online shopping stores (Flipkart, EBay, Amazon, etc.) to reduce fraudulent activity, time to time analysis of inventory, detect the shopping patterns of the customers, etc. There directly connected with the decision-making and profit-loss matters of the shopping industry.

6. **Government Sector and Insurance**

   Big data is being used by the food authority to identify and research disease and sickness trends that are related to food. This enables a quicker response, which has resulted in quicker treatment and fewer fatalities. For security purposes (including border security), big data technologies are extensively used by various government agencies and military forces. Big data also enables insurance businesses to better retain their customers through analyzing the pattern and behavior of their existing customers from historical records collected from social media, CCTV footage, CIBIL score, and fraudulent activity.

7. **Manufacturing and Energy Sector**

   Big data makes it possible to use predictive modeling to help decision-making after absorbing and combining significant amounts of text, temporal, graphical, and geospatial data. There are a lot of challenges the manufacturing industry facing today. Those challenges can be efficiently handled by big data technology.

2.3 Cloud Computing: Definition, Models, and Architectures

Inspired by the grid computing and utility computing movements, cloud computing appears and handles hardware and software resources efficiently from a large data center via the high speed Internet. The customer can pay as per their use of computing, storage, and communication resources. The terms "cloud" represents Internet and "computing" refers to the processing on those various resources of Internet. The cloud computing concept is based on one single question, "Why we purchase resources if we can rent them?" Therefore, cloud computing can be defined as Internet-based computing where on-demand (pay as you go) access from a collection of resources are accomplished without strong intervention of the service provider [5, 6].

Case Study-1

European researchers switched from supercomputers to cloud computing. High-performance computing (HPC) takes the help of powerful computers to solve high-end complex problems and that generates high-wage jobs. On average, 95% of

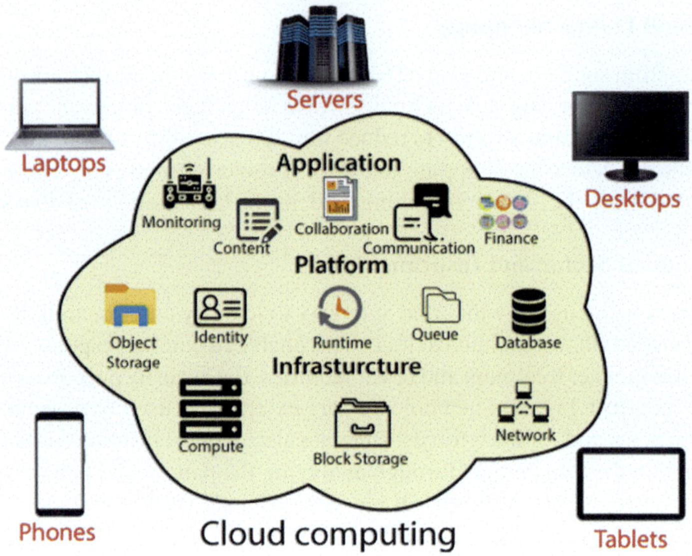

Fig. 2.4 Overview of cloud computing

the computing capacity of desktop computers in universities is wasted. By using Windows Azure, this capacity is maximized up to 99%. The traditional European supercomputing industry has largely vanished. So, the main objective to adopt cloud computing here is to better resource utilization. An overview of a cloud computing system is shown in Fig. 2.4.

Case Study-2 (Amazon Web Services)
Actual Analytics creates automatic, aided video content analysis tools that make it possible to index and search video content based on what is happening in the video. Their products make it possible to identify illnesses and medication side effects for usage in the pharmaceutical sector. The business made the early decision to employ a cloud platform to deploy their application because of the significant and fluctuating processing needs associated with video processing. So, the main objective to adopt cloud computing here is to provide rapid elasticity.

Case Study-3 (Microsoft Azure)
Anyone in the fishing sector should be extremely concerned about the high incidence of maritime fatalities, as 24,000 crew members are thought to drown globally each year. Both man overboard (MOB) Guardian and GeoPoint utilize personal safety equipment that can automatically set off on-vessel alarms while transmitting a signal via satellite to the search-and-rescue agency in the event of a man overboard. So, the no capital expenditure objective is adopted through cloud computing here. By the use of this cloud technology, numerous numbers of lives were saved at sea in UK.

Case Study-4 (Email Server Solution)

An organization requires an email server.

- Server computer.
- Operating system suitable for the server.
- Software (Microsoft Exchange Server version).
- Manpower for possible maintenance and updation.

Flip Side

- Software may only last for three years, whereas a server may last for five. Consequently, a compatibility gap may exist.
- Fault-tolerant issues can be faced by the server.
- If we do not anticipate rate of growth, peak size, etc., progress will be very challenging.
- Investment for necessary purposes.

Cloud Solution

- Set up your mail server using Gmail.

Advantages

- Maintenance is not required.
- There is no latency, and fault tolerance is assured.
- Scalability to infinity.
- Necessary features.
- Minimal lock-in.

Essential Characteristics

- Resource pooling and elasticity.
- Shared infrastructure.
- Uses recent advances in technology.
- Centralized and homogeneous.
- Cost-effective and on-demand self-service.
- Maintenance and security.
- Virtualization.

2.4 Comparison of Cloud with Other Computing

2.4.1 Cloud Versus Grid Versus Utility Computing

- A computing grid is a distributed system consisting of a large number of loosely coupled, heterogeneous, and geographically dispersed systems in different administrative domains.
- Heterogeneous versus homogeneous.
- Decentralized versus centralized.
- Utility computing refers to the ability to charge the offered services, and charge customers for exact usage, cloud computing also works like utility computing, you pay only for what you use but cloud computing might be cheaper, as such, cloud-based app can be up and running in days or weeks.
- Utility computing is a good choice for less resource demanding, whereas cloud computing is a good choice for high resource demanding.
- Utility computing users want to be in control of the geographical location of the infrastructure, whereas in cloud computing, provider is in complete control of cloud computing services and infrastructure.

2.4.2 Cloud Models

Cloud models are broadly classified into three types, service models, deployment models, and cloud cube model.

A. **Service Models**

Any cloud computing service model is built with the combination of Software as a Service (SaaS), Platform as a Service (PaaS), and Infrastructure as a Service (IaaS). Each model is described below,

A.1 **Software as a Service (SaaS)**

SaaS model is a model where user can directly use the software provided by a third part using pay-as-you-go basis. No up-front commitment is required here from the customer. Customer can use this service against a long-term contract. All these applications are running on the third-party provider's platform. As for example, exchange online is a full-featured email, calendar, and contacts solution delivered as a service hosted by Microsoft.

A.2 **Platform as a Service (PaaS)**

PaaS is a model, which outsources basic infrastructure and platform, such as Windows, UNIX, Solaris. PaaS facilitates deploying applications without the cost and complexity of buying and managing the underlying hardware and software where the applications are hosted. The customer uses their own applications. You may launch and manage apps more quickly in the AWS cloud by using AWS Elastic Beanstalk. Elastic Beanstalk will take care of the deployment specifics, including capacity provisioning, load balancing, auto-scaling, and application health management, once you quickly submit your application. Figure 2.5 shows the three layers hierarchy of cloud service model.

A.3 **Infrastructure as a Service (IaaS)**

IaaS gives the opportunity to access ground level resources such as physical machines, virtual machines, storage, networking components, servers.

Case Study Representing Service Models

You are able to prepare meals for four people at home. Twenty individuals are coming in with you.

- You place a Domino's pizza order (SaaS).
- If you pay your neighbor by the hour, she will lend you her gas burner and gas cylinder so you may prepare food for four people (PaaS).
- You ask permission to use the community hall, where you can prepare meals, from the chairman of your housing committee (IaaS).

B. **Deployment Models**

The deployment models define the location of the cloud and how we can access it. There are four kinds of deployment models, private, community, public, and hybrid clouds.

- **Private Cloud**—The infrastructure is purely for the benefit of an organization; it may be placed on or off the organization's property, and it may be managed by the organization or by a third party.
- **Community Cloud**—Several groups share the infrastructure, which serves a specific community with common concerns (e.g., mission, security requirements, policy, and compliance considerations). It can be handled by the organizations themselves or by a third party, and it can be on-site or off-site.
- **Public Cloud**—The infrastructure is controlled by a company that sells cloud services and is made available to the general public or a major industry group.

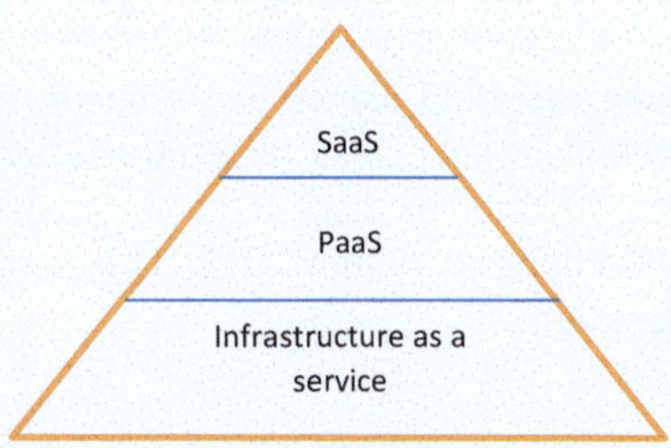

Fig. 2.5 Three-layered structure of service models

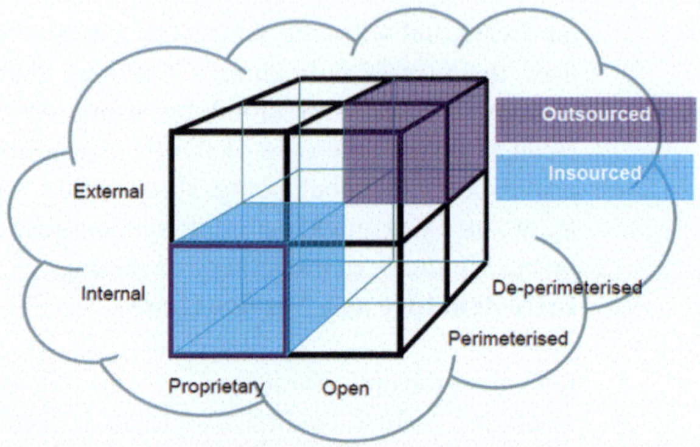

Fig. 2.6 Cubic model structure of cloud

- **Hybrid Cloud**—The infrastructure is made up of two or more clouds (private, community, or public) that are separate yet linked by standardized or proprietary technology that allows data and application portability (e.g., cloud bursting for load balancing between clouds).

C. **Cloud Cube Model**

The Cloud Cube concept aids in the categorization of cloud networks based on four dimensions. Their main goal is to keep the cloud network safe and secure. This cloud cube model aids in cloud formation selection for safe communication. Figure 2.6 shows an example of a cloud cube model.

2.4.3 Cloud Architecture

The entire cloud architecture depends on the following model and entities are shown in Fig. 2.7,

In this chapter, we have discussed three popular cloud platforms in brief [7, 8]. Besides that, we have also discussed an open-source private cloud platform and its basic architecture.

2.4.3.1 Amazon Web Services (AWS) Architecture

Amazon cloud named as "Amazon web services (AWS)" released its beta version around the year 2000. It has more upgraded version known as elastic cloud computing (EC2). With its IaaS delivery model services, AWS occupied the spotlight over 200 countries within 2012. This shows the international appeal of this computing paradigm.

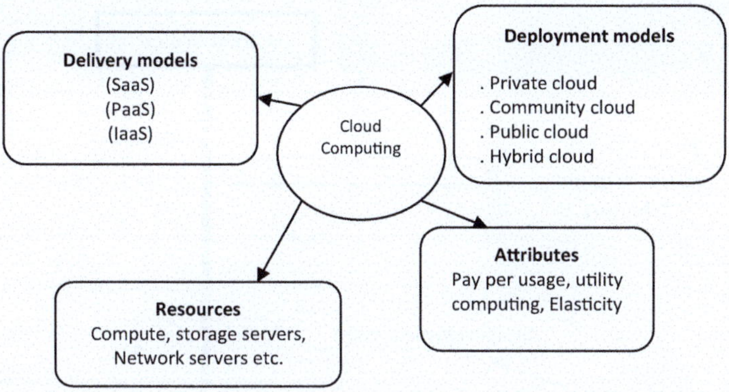

Fig. 2.7 Overview of cloud architecture

AWS has some notable advantages that makes it more popular.

a. Flexibility: In order to serve the diverse business needs of different organizations, AWS allows to use their familiar operating systems (OS), database, architectures, programming models, etc.
b. AWS provides a cost-effective cloud service without any long-term investment.
c. In terms of load balancing, scalability and elasticity is one of the major concerns which can be efficiently utilized by AWS.
d. AWS is capable to provide end to end security with a detail manual.

EC2 is mainly responsible to launch multiple applications under multiple VMs with different kind of operating systems. Those EC2 instances are communicated with the help of simple queue service (SQS). Simple storage services (S3), elastic block storage (EBS), and simple DB are various kinds of storage services available in AWS. The overall performance monitoring is supported by the cloud watch module. The auto-scaling (described in Sect. 5.2) is used to support elastic resource management. The parallel applications are directly migrated with the help of virtual private cloud (VPC). The entire services of AWS are handling and monitoring by AWS management control unit. Figure 2.8 shows an overall architecture of Amazon AWS cloud.

2.4.3.2 Google Cloud Platform

A group of cloud computing services are offered by Google under the moniker Google Cloud Platform (GCP). It is a platform for public cloud computing that offers a range of services, including compute, storage, networking, application development, big data, and more. These services are powered by the same cloud infrastructure that Google uses internally for its end-user products, including Google Search, Photos, Gmail, and YouTube, among others. GCP is a fully featured cloud platform that

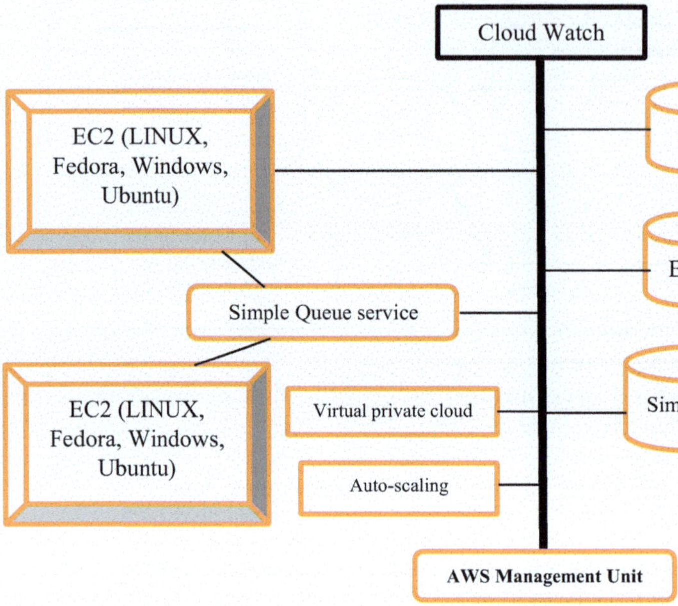

Fig. 2.8 AWS architecture

includes effective management of resources through proper scaling, multilevel security for resources, suitable network infrastructure (switches, wiring, routers, load-balancers, firewalls, etc.), and handling peak load through suitable bandwidth. Google offers a sizable number of services, each with a few special features. Because of this, Google Cloud Platform is constantly growing globally. Some of the useful services are listed below,

- Compute services.
- Networking.
- Storage services.
- Big data.
- Security and identity management.
- Management tools.
- Cloud AI.
- Internet of Things (IoT).

2.4.3.3 Microsoft Azure Cloud Platform

Microsoft's PaaS (Platforms as a Service) and SaaS (Software as a Service) cloud platforms are referred to as Azure and Online Services, respectively. An operating system is called Windows Azure, a cloud-based SQL Server is called SQL Azure, and a group of cloud application services is called Azure AppFabric (formerly known as

.NET Services). There are three main parts: compute, which offers an environment for computing, and storage, which offers scalable storage. It connects nodes made up of servers, fast connections, and switches using a fabric controller that launches, maintains, and monitors applications. Windows Azure's components are: Compute, which powers cloud applications; Blobs, tables, and queues are used in storage to hold data; applications are deployed, managed, and monitored by fabric controller; data cache copies are kept by content delivery network (CDN). The key components of Azure architecture are compute, storage, database, security, integration services, content delivery network, monitoring, networking, mobile services, workflows, and web services. The different Azure cloud components are shown in Fig. 2.9.

A. **Azure as IaaS**

   Infrastructure as a Service (IaaS) is the core component of the cloud platform. IT managers use this Azure service for processing, storage, networks, and all other standard computer activities. Users may execute any software, making it one of the Azure subjects to study.

B. **Azure as PaaS**

   A computer platform known as PaaS consists of an operating system, an environment for running programming languages, a database, and/or web services. Application vendors and developers use this Azure service. This platform is offered to the client to enable software development and deployment, as suggested by its name. One of the fundamental ideas behind Azure is that it frees the customer from worrying about infrastructure and hardware so they can concentrate on developing applications. Additionally, it handles problems with servers, networking, and operating systems.

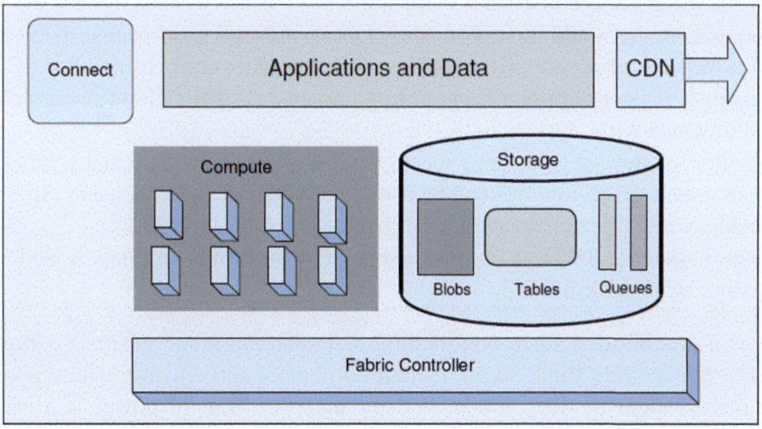

Fig. 2.9 Overall components of Microsoft Azure cloud

C. **Azure as SaaS**

Software as a Service (SaaS) is software that is hosted and controlled centrally. Each customer uses the same version of the application. Scaling out allows you to run numerous instances. This makes it easier for you to guarantee the greatest performance wherever. By way of a recurring monthly or yearly subscription, the software is licensed. Dynamics, MS Exchange, and Office are all available as SaaS.

2.4.3.4 Eucalyptus Private Cloud Platform

A private cloud's control system can be built using Eucalyptus, OpenNebula, and Nimbus. In this section, we have mainly focused on the discussion of Eucalyptus cloud architecture. The following steps are schematically performed by a cloud infrastructure.

Retrieves user input from the front-end; obtains user input from the front-end, finds a system, and requests the virtual machine monitor (VMM) running on that system to build a VM. It then launches DHCP22 and IP bridging software to assign the VM a MAC address and an IP address and obtains a virtual machine's (VM) disk image from a repository. The Eucalyptus cloud architecture along with threat detection system is represented in Fig. 2.10.

- The virtual machine Run under Xen, KVM, and Vmware, among other VMMs.
- The node controller runs on each server/node assigned to host a VM and manages the node's operations. A cluster controller receives reports.
- Controller for the cluster manages numerous servers interacts with each server's node controller to set the time for requests to be processed on that node. Cloud controller is in charge of cluster controllers.
- Cloud control gives administrators, developers, and end users access to the cloud. Both a web-based dashboard and command line tools compatible with EC2 can access it. It makes high-level scheduling choices, controls cloud resources, and communicates with cluster controllers.
- Controller for storage gives apps access to persistent virtual hard disks. It is EBS's correspondent. EBS volumes can be used to create snapshots by users. Snapshots are made accessible across availability zones and kept in Walrus.
- Storage assistance (Walrus) allows users to store things in buckets and offers persistent storage akin to S3.

This is a crucial idea since one method of managing resources in a computing cloud is to concentrate the load on a small number of servers and, where possible, put the rest of them in sleep mode. Performance per Watt of power is a measure of a system's operational efficiency; concentrating the workload on a small number

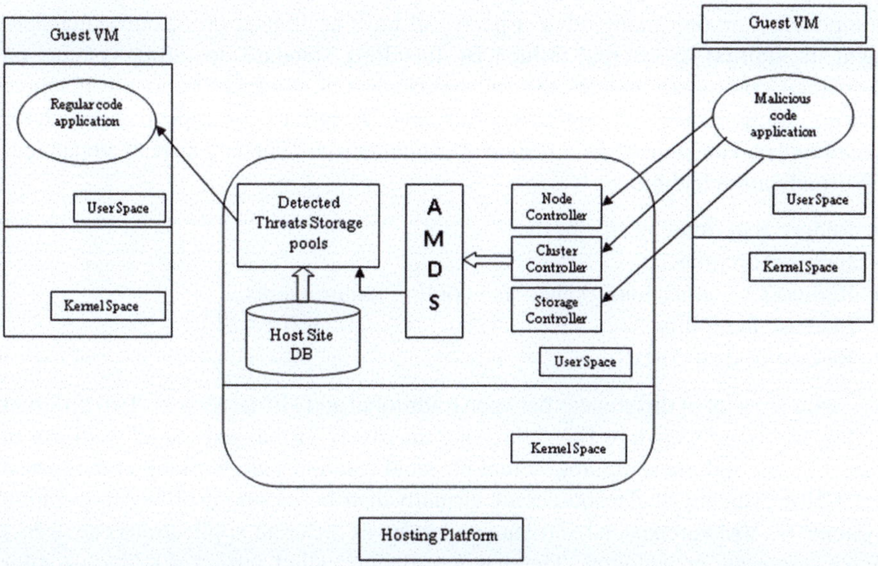

Fig. 2.10 Eucalyptus architecture with security detection block

of disks and allowing the others to run in a low-power mode is one way to reduce energy usage. Replication is a strategy that can be used to achieve this. Replication strategies are decreased by up to 31%. This leads to the notion of "Green Cloud". A lot of research works are going on this energy efficient cloud computing.

2.5 Load Balancing and Virtualization in Cloud Computing

2.5.1 Virtualization

Virtualization in cloud computing is a technique to map logical names into physical resources and share a single physical resource among many organizations. It is nothing but a creation of a logical version of some physical instances such as storage devices, hardware systems, network resources. The concept of virtualization is very similar with the concept of virtual memory in operating system and CPU sharing [9].

Motivation

It is highly challenging to manage resources for a group of users using a variety of programs that are running on various operating systems. When resources are overused and users are recalcitrant, resource management becomes even more difficult. The heterogeneity of the hardware and software systems, the capacity to approximate the system's overall status and redistribute the load, the failure rates of various components, among many other aspects, all have an impact on resource management in addition to external influences. Installing standard operating systems on individual computers and relying on traditional OS strategies to assure resource sharing, application protection, and performance isolation is the typical approach for a data center. The alternative is resource virtualization. There are a set of advantages of virtualization technique,

- Reduce real estate needs.
- Increase up time.
- Reduce CO_2 emissions, power, and cooling requirements.
- Increase flexibility.
- Reduce overall costs.

There are a lot of differences between traditional and virtual servers. Two different architectures are shown in Fig. 2.11. One massively virtualized model is shown in Fig. 2.12. In both these figures, actual physical resources are working at the lowest level. However, physical resources are virtually distributed among different operating systems by VMware special software and users of different applications can access those resources in isolation. There is a software called virtual machine monitor (VMM) whose main job is to divide and distribute the computer resources into one or more virtual machines (VMs). Under the control of VMM, a guest operating system (OS) is active in user mode. High-end security is provided by VMM with rigorous isolation. The VMM buffers interruptions that a guest OS suppresses until the guest OS activates them. Each guest OS has a shadow page table that is maintained by the VMM, which copies any changes made by the guest OS into its own shadow page table. This shadow page table points to the actual page frame and is used by the hardware component known as the Memory Management Unit for dynamic address translation (MMU) [10]. Figure 2.13 represents the types of virtual machines.

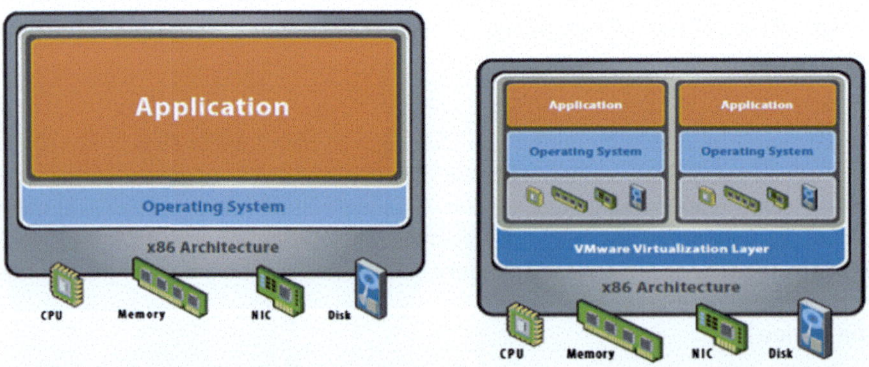

Fig. 2.11 Traditional versus virtual servers

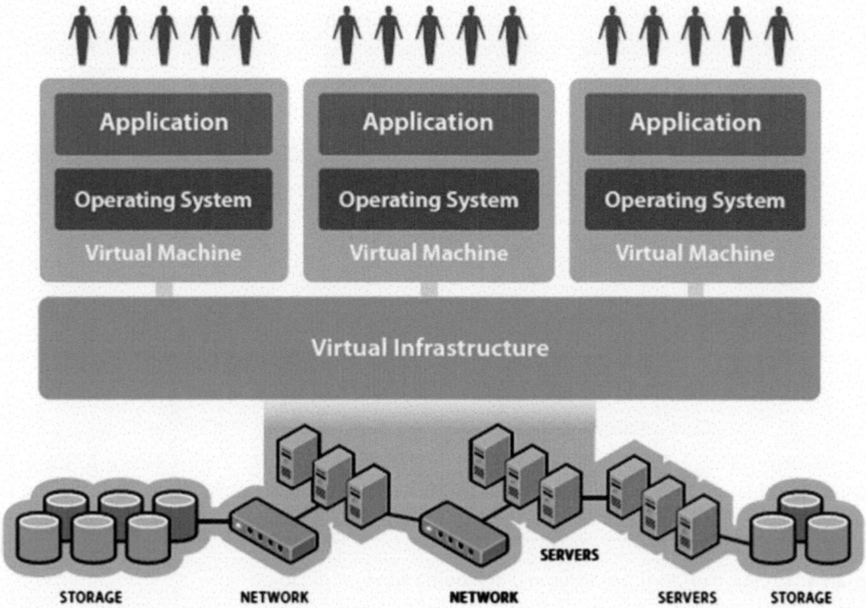

Fig. 2.12 Massive virtual cloud server model

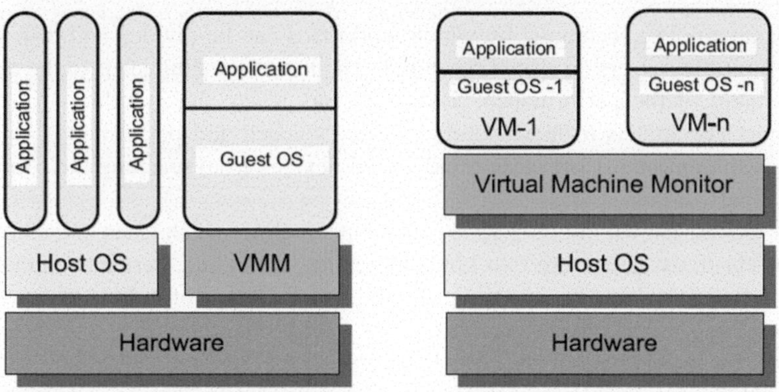

Fig. 2.13 Types of virtual machines

Figure 2.14 shows how machine images (AMI instances) are created in a virtual cloud environment of Amazon EC2 architecture and how services are deployed over those instances in cloud. It describes that N1, N2, and N3 are virtual instances which are created on the physical architecture lower backbone (LB). Therefore, three services are deployed and executing over those three instances.

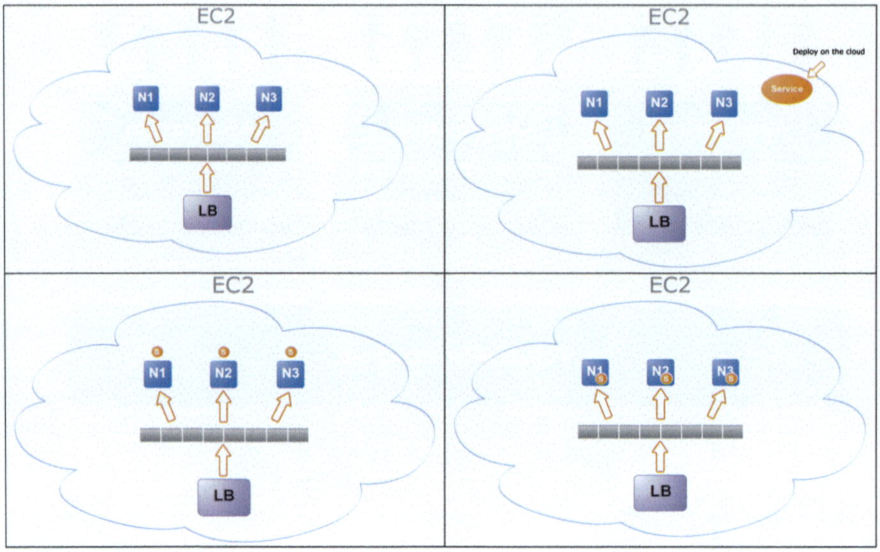

Fig. 2.14 Creation of AMI instances and deploying services in cloud

2.5.2 Load Balancing

Load balancing is a technique by which workloads can be divided and distributed among multiple servers or virtual machines. The load balancing leads to the overall improvement of the performance. Several factors led to the introduction of load balancing. The first is to increase each device's speed and performance, and the second is to continually preventing individual devices from reaching their limit by reducing performance [11].

This process shown in Fig. 2.15 is called auto-scaling which leads to the concept of rapid elasticity. There are two kinds of scaling technique, vertical scaling, and horizontal scaling. In vertical scaling, we increase the loads of individual resources (memory, processor, hard disk drive, etc.) of a VM into a certain extent instead of spreading the load across all the VMs. In horizontal scaling, if more traffic is coming, then load balancer added new virtual machines dynamically. It is also adding virtual machines parallely [5, 12].

Figure 2.16 shows an example of auto-scaling over Amazon EC2 architecture. In Fig. 2.16, how load increases with the new instances and how load decreases after terminating those instances are shown. Auto-scaling plays a very vital role in this situation while handling the increasing and decreasing of loads. However, there are various load balancing algorithms suitable for cloud environment [13].

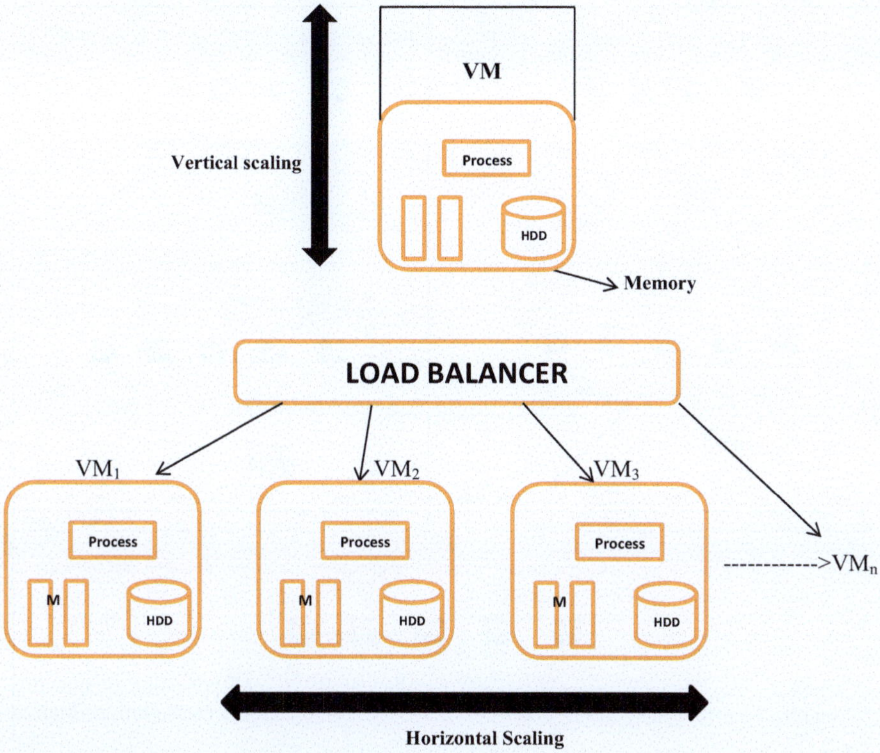

Fig. 2.15 Autoscaling in cloud load balancing

A. **Static Algorithm**

Static algorithms are designed for systems with extremely little load variance. The servers in the static algorithm share the whole traffic equally. For this algorithm to run more efficiently on processors, which is decided at the start of implementation, it requires a deep knowledge of server resources. However, the choice to move the load is independent of the system's current condition. One significant disadvantage of this algorithm is that additional machines for load balancing cannot be applied.

B. **Dynamic Algorithm**

The load balancing preference is first given via dynamic algorithm to the network's lightweight server. Here, the load is managed by the system's current state. Making load transfer choices based on the actual condition of the system at the time is the expertise of dynamic algorithms. In this system, processes are permitted to switch in real time between heavily and sparsely used machines.

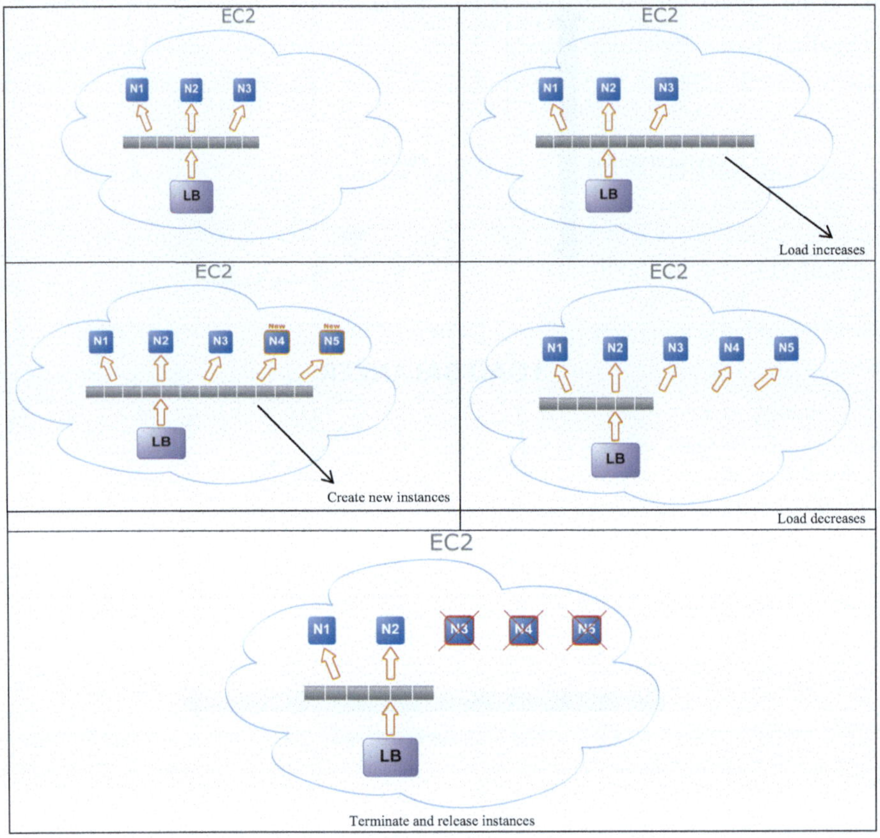

Fig. 2.16 Auto-scaling over Amazon EC2 architecture

C. **Round Robin Algorithm**

This algorithm assigns the work using the Round Robin method, as the name would imply. In the beginning, it randomly selects the first node and rounds robin distributes the work to the remaining nodes. Processors assign each process in a cyclical manner without specifying a priority. When the workload allocation between the processes is comparable, this leads to a quicker response. Every process has a unique loading time. As a result, certain nodes may be overloaded while others may continue to be underutilized.

D. **Weighted Round Robin Algorithm**

The most difficult problems with the Round Robin algorithm are improved by the weighted approach. Jobs are allocated according to the weight values, which has a predetermined fixed weight. Greater capabilities among processors are given a higher value. As a result, servers with higher weights will receive more tasks. Server traffic will remain constant after all weights are equal.

E. **Opportunistic Load Balancing Algorithm**

It is mainly useful to maintain node the activity of nodes. The present burden of each system is never taken into account. This algorithm randomly assigns all open jobs to each of these nodes, regardless of the workload that each node is currently carrying. The processing work will be carried out slowly because this algorithm does not account for the node's implementation timing, which causes some bottlenecks to develop even though some nodes are available.

F. **Minimum to Minimum Load Balancing Algorithm**

In this algorithm, the jobs that take the least amount of time to complete are first identified. A minimal value is chosen among all the tasks out of those. The job is scheduled on the computer according to that minimum amount of time. The other jobs on the machine have been updated similarly, and the job is then taken off of that list. This process will keep on until the last job is given. This algorithm appears to work best when a large number of smaller jobs prevail over large tasks.

G. **Maximum to Minimum Load Balancing Algorithm**

It resembles the previous one almost exactly. However, this algorithm differs slightly from the previous one. Here, the maximum value is chosen after determining the shortest implementation time. The work is scheduled on the machine after that according to the maximum time period. All of the tasks' completion times are updated, and the given work is taken off the list.

All of the aforementioned load balancing techniques were important for your server cluster. After carefully considering everything, you can select the best option from the group. Avoid choosing the incorrect load balancer since it can negatively impact your system performance [8].

2.6 Cloud Computing Systems for Data-Intensive Applications

Cloud-based data-intensive applications are used to understand, handle, and analyze high volume of data with the help of virtual parallel computing. This kind of large amount of data is called big data whose size ranges from terabytes to petabytes generally. Data-intensive applications help to understand the complex problems timely and in a meaningful way. Bayesian analysis on bulky dataset in the cloud is an example. By creating new classes of software, hardware, and algorithms, data-intensive applications respond to the constantly increasing data complexity and accompanying analysis needs by delivering rapid and insightful analytical results. Through parallel processing pipeline technique, it minimizes the high volume of data into relatively small datasets but very rich and informative. Therefore, the human understands these small rich datasets easily. At the first phase of preprocessing, removing noise and summarization has been done on the raw data and useful knowledge has been fetched from the preprocessed data through some intelligent algorithms. For parallel processing, MapReduce can be used to distribute the data across the clusters comprises thousands of nodes and process them parallely at each node. The basic challenge in data-intensive computing is the incoming data from heterogeneous resources with diverse types, scales, and quality. In this scenario, sometimes one job requires data distributed in different data sources at the same time. To overcome this issue, cloud computing develops a complete Internet-based scientific workflow system. Unlike the grid computing, cloud computing provides an efficient, effective, fault-tolerant, reliable, less costly automatic workflow system. Through the Internet, users can access scientific cloud workflow systems from anywhere in the globe and upload their data and run their programs. It is simple to exchange and distribute the data across users because all of the data are controlled in the cloud. For data-intensive jobs, the combination of hybrid cloud is appropriate. A hybrid cloud is made of private and public cloud. Private cloud provides a secure environment to handle user resources securely and public cloud provides scalability of the user resources along with the "pay-as-you-go" scheme. This hybrid cloud system provides a perfect balance between performance (time) and cost for this kind of jobs. The input datasets are mirrored among sites with varying computing and data storage capabilities with the notion that the majority of data-intensive applications require a sizable number of input datasets [14].

2.7 Analytical and Perspective Approach of Big Data in Cloud Computing

Big data applications entail the storing, analysis, and processing of enormous volumes of data using sophisticated computing systems. Unfortunately, the needs of such data and applications cannot be met by conventional computer and storage technologies. Another issue is the requirement for restricted and controlled data sharing among researchers in various places. Along with it, bandwidth is also necessary for the data transfer. The majority of these issues have potential answers thanks to cloud computing [15, 16].

Many big IT firms have selected HDFS for big data storage solutions that require high configuration of the hardware. The general cloud storages (Google Drive, iCloud, etc.) are used to store such big data and overcome the limitations. The potential uses of big data have been further increased by cloud computing. Developers may easily create ad hoc clusters in the cloud to test a small fraction of data since it enables genuinely elastic scalability.

The relationship between big data and cloud computing is shown in Fig. 2.17. The input is a kind of big data in an organized or unorganized format that has been gathered from a variety of data sources, including cellular and other IoT devices. Using Hadoop or other data stores, this vast amount of data is cleaned before being

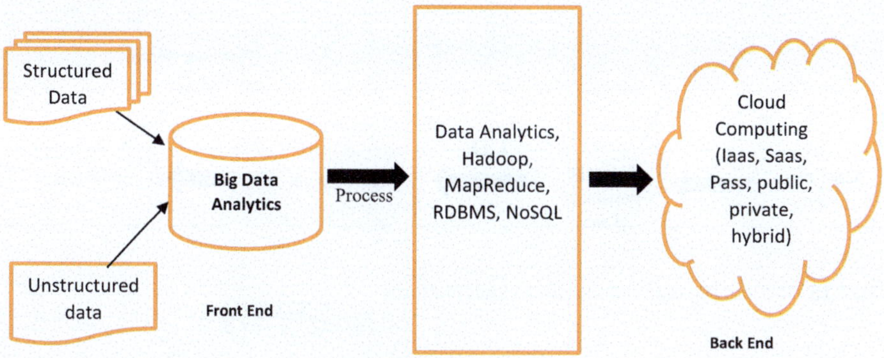

Fig. 2.17 Relation between big data and cloud computing

saved. In order to provide services, the saved data is processed again using cloud computing tools and techniques. All of the tasks necessary to transform input data are included in the processing phases. The value obtained following the processing of data for analysis and display is known as output. Both big data and cloud computing is tightly coupled through one element is called Internet of Things (IoT). Massive amounts of IoT device-generated data require immediate analysis. Cloud service providers keep data in data stores and enable data transmission via the Internet or via lease lines. The big data that has been saved in the cloud is then subjected to cleaning and analysis using cloud computing techniques and tools. It offers a mechanism for the data to travel, be stored, and be examined. Big data and IoT have a shared platform thanks to cloud computing. Big data, IoT, and cloud computing all work well together [17].

Big data as a service is a famous example of big data with cloud. For example, BigML is a machine learning platform built on the cloud that enables the creation of predictions for online big data. The Hadoop platform Cloudera utilizes the AWS cloud. MapR is a Hadoop-based platform that enables businesses to quickly save and handle big data partners and clients, such as Google and Amazon, have adopted it.

As shown in Fig. 2.18, IoT is the source of the data, and big data is an analytical technological platform for the data. The entire framework is running at SaaS cloud platform [18].

2.8 Conclusion

Cloud computing and big data are closely related with each other and has a huge number of applications in IT service industry. The use of cloud services changes the entire scenario of handling such huge amount of data efficiently and effectively. It has altered how information technology is used and made the on-demand service model's promises a reality. In this chapter, we describe big data technology, cloud

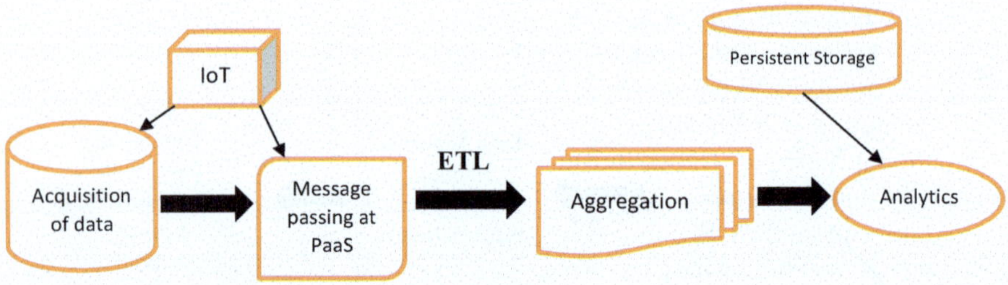

Fig. 2.18 IoT and big data running on a cloud platform

service models, various services and their respective applications in brief. To ensure the long-term success of data management in a cloud computing environment and to cooperatively explore new horizons, researchers, practitioners, and social science academics should work together.

References

1. Dey N, Hassanien AE, Bhatt C, Ashour A, Satapathy SC (eds) (2018) Internet of things and big data analytics toward next-generation intelligence, vol 35. Springer, Berlin
2. Sarowar MG, Kamal MS, Dey N (2019) Internet of things and its impacts in computing intelligence: a comprehensive review—IoT application for big data. In: Big data analytics for smart and connected cities, pp 103–136
3. Dey N, Das H, Naik B, Behera HS (eds) (2019) Big data analytics for intelligent healthcare management. Academic Press
4. Rajeswari P, Viswanadha Raju S, Ashour AS, Dey N (2017) Multi-fingerprint unimodel-based biometric authentication supporting cloud computing. In: Intelligent techniques in signal processing for multimedia security. Springer, Cham, pp 469–485
5. Mastelic T, Oleksiak A, Claussen H, Brandic I, Pierson JM, Vasilakos AV (2014) Cloud computing: survey on energy efficiency. ACM Comput Surv (CSUR) 47(2):1–36
6. Cao K, Hu S, Shi Y, Colombo AW, Karnouskos S, Li X (2021) A survey on edge and edge-cloud computing assisted cyber-physical systems. IEEE Trans Industr Inf 17(11):7806–7819
7. Kaushik P, Rao AM, Singh DP, Vashisht S, Gupta S (2021) Cloud computing and comparison based on service and performance between Amazon AWS, microsoft Azure, and google cloud. In: 2021 international conference on technological advancements and innovations (ICTAI). IEEE, pp 268–273
8. Gupta B, Mittal P, Mufti T (2021) A review on Amazon web service (AWS), microsoft Azure & google cloud platform (GCP) services
9. Mansouri Y, Babar MA (2021) A review of edge computing: features and resource virtualization. J Parall Distrib Comput 150:155–183
10. Sharma VK, Singh A, Jaya KR, Bairwa AK, Srivastava DK (2022) Introduction to virtualization in cloud computing. In: Machine learning and optimization models for optimization in cloud. Chapman and Hall/CRC, pp 1–14
11. Xu M, Tian W, Buyya R (2017) A survey on load balancing algorithms for virtual machines placement in cloud computing. Concurr Comput Pract Exp 29(12):e4123
12. Fazli A, Sayedi A, Shulman JD (2018) The effects of autoscaling in cloud computing. Manage Sci 64(11):5149–5163

13. Patra PK, Singh H, Singh R, Das S, Dey N, Victoria ADC (2016) Replication and resubmission based adaptive decision for fault tolerance in real time cloud computing: a new approach. Int J Serv Sci Manag Eng Technol (IJSSMET) 7(2):46–60
14. Mehdi NA, Holmes B, Mamat A, Subramaniam SK (2012) Sharing-aware intercloud scheduler for data-intensive jobs. In: 2012 international conference on cloud computing technologies, applications and management (ICCCTAM), pp 22–26
15. Zanoon N, Al-Haj A, Khwaldeh SM (2017) Cloud computing and big data is there a relation between the two: a study. Int J Appl Eng Res 12(17):6970–6982
16. Elazhary H (2014) Cloud computing for big data. MAGNT Res Rep 2(4):135–144
17. Pal R, Anand T, Dubey SK (2018) Analytical and perspective approach of big data in cloud computing. In: Big data analytics. Springer, Singapore, pp 233–239
18. Ullah S, Awan MD, Sikander Hayat Khiyal M (2018) Big data in cloud computing: a resource management perspective. In: Scientific programming

Chapter 3
Edge Computing with Internet of Things (IoT) and Data Analysis

3.1 Introduction

A distributed computing paradigm known as edge computing architecture moves data storage space nearer to the data's original sources. Response times should increase, and bandwidth should be reduced as a result. Instead of transmitting IoT data back to a warehouse or cloud, it enables edge processing of IoT data [1]. IoT and edge computing work well together to quickly analyze data in real time. IoT is simply working as a bridge between digital and physical worlds with the help of a set of sensors and actuators. Internet plays a vital role in communication among the day-to-day useful devices ("things"). The storage, processing, analysis, and transportation of data produced by thousands of IoT devices are being transformed by edge computing. IoT has various applications in the real world. The Internet of Things connects common objects equipped with electronics, software, and sensors to the Internet, allowing for the collection and sharing of data without the need for human contact (IoT) [2]. For example, air-conditioners automatically adjusted its temperature (cooling system) based on the humidity and outside heat. As an illustration, consider a household where all of the appliances, like as the air conditioner and lights, can be connected to one another and controlled from the same platform. We have a platform that allows us to link our car, monitor its fuel gauge and speed, and track its whereabouts. It would be amazing if there was a central hub where everything could link to each other so that I could control the temperature of the room according to my preferences. For instance, if I prefer my home's temperature to be set at 18 or 22 °C when I return from work, my car's location-based AC system would begin operating well in advance of my arrival at home [3].

Figure 3.1 shows a basic communication of home appliances with IoT devices through Internet cloud. There are some important components of an IoT system.

A. **Sensors**

A crucial element that enables you to gather real-time data from the outside world is a sensor or device. There could be different degrees of complexity in

S. Chakraborty and L. Dey, *Computing for Data Analysis: Theory and Practices*, Data-Intensive Research, https://doi.org/10.1007/978-981-19-8004-6_3

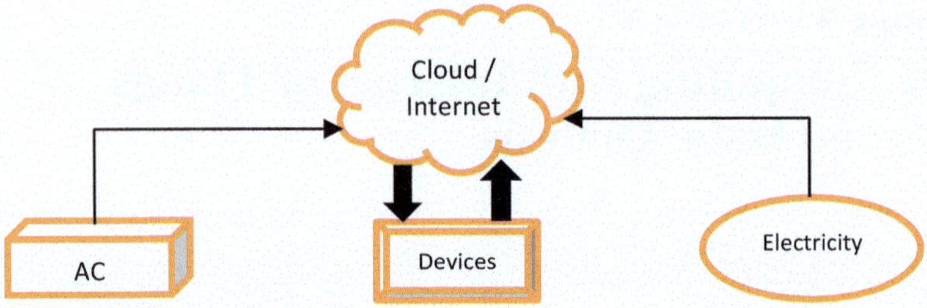

Fig. 3.1 Communication at IoT platform

all this data. It might only be a sensor for measuring the temperature, or it might be a video stream. A gadget may feature a variety of sensors that are used for more than just sensing. A cell phone, for instance, has a number of sensors, such a GPS and a camera, but your smartphone is unable to feel these things [4].

B. **Communication**

A cloud infrastructure receives all the collected data. The sensors ought to be linked to the cloud utilizing a variety of communication channels. These communication channels include WANs, Bluetooth, WiFi, mobile or satellite networks, etc.

C. **Processing of data**

The software processes the gathered data once it has been captured and transferred to the cloud. Simply checking the temperature might be considered part of this process. However, there are situations when it can also be quite difficult, such as when utilizing computer vision on video to identify things.

D. **Interfacing with Users**

The end-user has to have access to the information, which can be done by setting alarms on their phones or notifying them by email or text message. The user may occasionally require an interface that continuously monitors their IoT equipment. The user, for instance, has a camera set up in his house. He wants to use a web server to view all of the streams and video recordings.

3.2 Related Technologies, Architectures, and Protocols of IoT

The architecture of IoT system has been changing since last decade where different researchers provided different architectures. IoT system is mainly classified into three- or five-layer architecture [5].

3.2.1 IoT Architectures

A. **Three-layered IoT Architecture**

The three-layered architecture is divided into perception layer, network layer, and application layer [6].

- The **perception layer** can be compared with the physical layer of the OSI model. It has sensors that sense the environment and collect data about it. It detects certain physical factors or locates other intelligent devices in the surrounding area.
- The **network layer** is in charge of establishing connections with other intelligent objects, network components, and servers. Its capabilities are also employed for processing and transferring sensor data.
- Providing application-specific services to the user is the responsibility of the **application layer**. It describes a variety of uses for the IoT technology, including smart homes, smart cities, and smart health.

The three-layer design encapsulates the core concept of the Internet of Things; however, research on IoT frequently focuses on its more intricate details; therefore, it is insufficient. Therefore, the five-layered architecture was introduced for the sake of research community, and it includes transport layer, processing layer, and business layer [7].

B. **Five-layered IoT Architecture**

The first three layers are working in a similar fashion.

- Through various kinds of networks (LAN, WiFi, etc.), **the transport layer** moves sensor data from the perception layer to the processing layer and back again.
- The middleware layer is another name for the **processing layer**. It takes in a ton of data from the transport layer and processes, saves, and analyzes it. It is capable of managing and offering a wide range of services to the lower tiers. It makes use of a variety of technologies, including big data processing modules, cloud computing, and databases.
- The **business layer** oversees the whole Internet of Things (IoT) system, including all apps, revenue streams, and user privacy. The business layer is not covered in this essay. So, we stop talking about it here (Fig. 3.2).

C. **Basic Components in IoT Architecture**

In a typical social IoT scenario, the devices and services are treated as bots that may establish interactions and change those relationships over time. This will enable us to easily let the gadgets communicate with one another to cooperate and complete a challenging assignment [6, 8].

Fig. 3.2 Five-layered IoT architecture

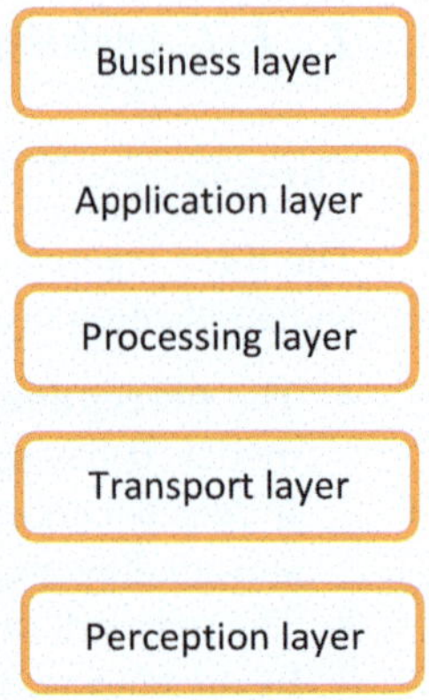

i. **Types of Sensors and Actuators**

IoT sensors are often compact, inexpensive, and power-efficient. Things like battery performance and deployment ease limit them.

(a) **Mobile phone sensors**

In general, our smart phones embedded with various kinds of sensors. Many researches are going on to ensure the flexibility, security, and reliability of smart phones. On the smartphone, applications can be created that leverage sensor data to deliver valuable results. A mobile phone's accelerometer detects motion and speed. The smartphone's variations in velocity are normally measured in three dimensions. Accelerometers come in a variety of velocities [9]. The accelerometer's data patterns can be used to identify the user's physical tasks, such as bicycling, driving, and running. The orientation changes of your smart phone are measured by gyroscope. Since they record both visual and auditory data, the camera and microphone are particularly potent devices that can be used to detect many kinds of circumstantial data based on customer's current environmental communication. Technologies like recognition of voice and acoustic features can be used to interpret the audio data. The magnetometer identifies the presence of metals through its magnetic field detection capability. One of the most important information is the tracking of the mobile phone location through global positioning system (GPS). It measures the distance from a number of satellites (or mobile phone towers) and computes the coordinates. The light sensor measures the surrounding light intensity. It can be used to adjust the screen's brightness and the others whose actions rely on the level of ambient light. For instance, we can adjust the lighting in a space. The infrared (IR) LED used by the proximity sensor generates IR rays. When they hit an object, these rays return. We can determine the distance based on the time difference. This makes it possible to calculate the distance between various objects and the phone. Some smartphones have feature additional sensors to estimate the heat, humidity, respectively. Numerous applications for cell phones that track a person's health and fitness are currently being developed. They monitor users' physical activity, nutrition, exercise, and lifestyle to assess their fitness level and make recommendations to them in line with that assessment [10].

(b) **Sensors for healthcare**

The IoT and its various sensors can be very beneficial for various applications in health care industry. These applications may be used to monitor the health of the patients. They can then give the doctor, the patient's family, or both real-time comments. The market is filled with wearable sensor technology. They have medical sensors built in that can measure a variety of factors, including high or low pressure, temperature of body, heart rate, and blood sugar levels [11–13].

Furthermore, thanks to the introduction of cutting-edge features by businesses like Apple, Samsung, and Sony, intellectual watches and fitness tracking device are growing in popularity on the business market. For instance, a smart watch (Fig. 3.3a) has capabilities like smartphone connectivity, accelerometer sensors, and heart rate monitors. Electronic rubbery skin-based patches are used to monitor health and transmit the health data in a wireless mode. This is one of the novel discoveries in IoT devices. Similar to tattoos are monitoring patches. They are disposable, stretchy, and quite affordable. These patches help the patients to keep track their health issues. Neurosensors (Fig. 3.3b) are mainly used to collect different frequencies in the form of EEG brain signals. It is possible to determine if the brain is at ease or racing with ideas based on these types of frequencies of signals [14, 15].

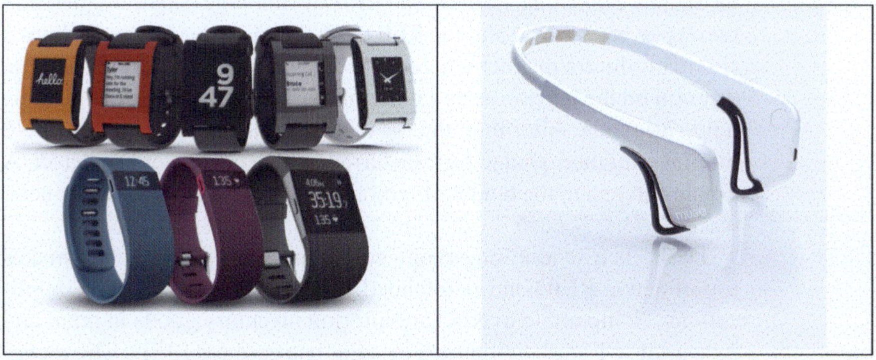

(a) (b)

Fig. 3.3 **a** Smart watches with healthcare sensors and **b** neurosensory-based headband for detecting brain signals

(c) **Sensors for environmental activity**

Environmental sensors sense the environment's properties, such as air pollution and water pollution. A thermometer and a barometer are tools used to measure several parameters, including pressure and temperature. Although it can also contain air quality, moisture, volatile organic compounds (VOCs), and seismic sensors, temperature sensors are among the most used environmental sensors. Besides that, there are some other sensors such as, gas sensors, rainfall sensors, sensors measuring the intensity of light, soil sensors, wind speed and direction sensors, and so on.

(d) **Sensors for measuring chemical substances**

These sensors can be used to assess food related items and agricultural items, keep an eye on food standards in smart intelligent kitchens, and monitor pollution levels in smart cities. As for example, in a variety of applications, including the detection of combustible gases in industrial plants and ethanol in breathalyzer tests, gas sensors and detectors are applicable. The most popular gas sensors detect a range of gases, from butane to hydrogen, and are based on metal oxides [16].

(e) **Radio Frequency Identification (RFID)**

RFID is an identifying system in which data is stored on an RFID tag, which is a tiny chip with an antenna, and read, by an RFID reader. The tag uses radio waves to convey the information it contains. It resembles bar code technology in certain ways. However, unlike a conventional bar code, it can recognize itself from a distance even without a human intervention and does not need line of sight connection. The frequency affects the RFID's range. It can travel up to many hundred meters. RFID is classified into two types of tags, active and passive. The reader's electromagnetic waves provide electricity for passive tags, making them inexpensive and long-lasting. The RFID technology can be divided into two types, near RFID and far RFID. An RFID reader nearby utilizes a coil to produce a magnetic field and conduct switching electricity. The tiny coil on the tag produces a potential as a result of ambient magnetic field variations. After that, the tag can generate a weak magnetic field that encodes the signal to be transferred, which the reader can detect. A dipole antenna in the reader of remote RFID transmits electromagnetic waves [17].

Dash7 is a secure long-range communication protocol that makes use of active RFID and is intended for use in commercial IoT applications. Comparable to NFC, monitoring inventory goods in retail and industrial IoT applications is a popular use case for RFID. The movement of objects can also be tracked by RFID. For instance, tiny chips are sometimes bonded to the front of cars. When the vehicle approaches a roadblock with a reader, it examines the data on the tag to determine whether the vehicle is authorized. If so, it opens on its own. People are given RFID cards, which enable them to be recognized by an RFID reader and granted access as necessary.

(f) **Actuators**

An actuator is a tool that can alter the surroundings by transforming electrical energy into a usable type of energy. Some loudspeakers, lighting, heating or cooling components, motors are some examples. A complex example of an actuator used in the Internet of Things is a digital finger, which is applied for switch on and off movements (or other devices that require slight movements) in wireless mode. Consider a system for a smart house, which includes numerous sensors and actuators. Lock and unlock doors, turn on and off LEDs and other electronic equipment, warn customers of potential risks through alarms or alerts, and regulate a home's temperature through actuators. Electrical, hydraulic, and pneumatic are three kinds of actuators used for various tasks.

Apart from these, the bottom device layer can be consisted of other useful directly connected processing devices. For example, Ethernet or WiFi-based Arduino and Raspberry Pi, ZigBee devices connected through a ZigBee gateway, mobile phone-based Bluetooth devices, etc. It is advised that each device have an EEPROM-stored OAuth2 Refresh and Bearer token as well as a UUID (ideally an immutable ID generated by the core hardware). The specification is based on HTTP, but the reference design also allows these flows over MQTT (as we shall explore in the communications section) [18].

ii. **Data Preprocessing**

The sensors data have been collected, processed, analyzed in cloud platform as it offers massive data handling, scalability, and flexibility. However, the use of cloud framework for IoT applications is not suitable always.

The reasons are listed below,

- The majority of smart devices are portable. Due to shifting network circumstances in various areas, communication with the cloud data center is challenging when their position changes. This mobility issue can be removed by introducing a very new mobile cloud computing service.
- It takes time to communicate with the cloud and receive responses. Applications that depend on real-time answers and are latency sensitive might not be possible with this paradigm. Additionally, lossy wireless networks during communication can result in inaccurate data.
- Latency can be increased.
- IoT gadgets run-on batteries, and connectivity uses a lot of power. As a result, they cannot afford constant communication.

To overcome all these issues, fog computing can be employed along with the cloud. A cloud that is near the ground could be used to represent the

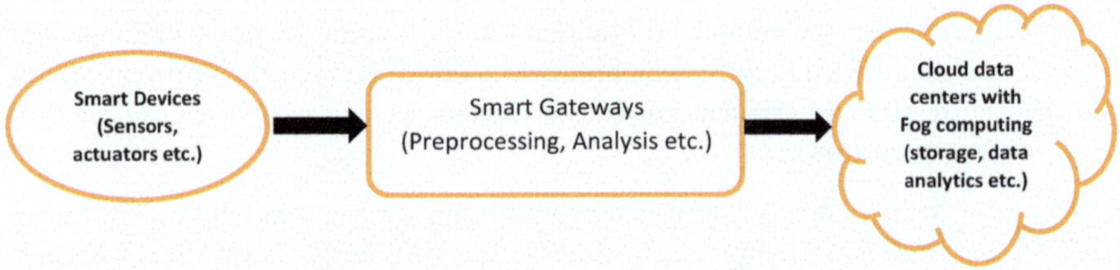

Fig. 3.4 Data preprocessing steps

fog. The cloud and fog paradigms are complementary. For IoT applications to run at their best, both are necessary.

In Fig. 3.4, smart gateway gathers data from sensors, preprocesses and filters those data, offers computing platform, storage capacity, and networking services to IoT devices, communicates with the cloud and sends only required information, monitors IoT device power consumption, keeps tabs on their operations and services, and ensures data security and privacy.

iii. **Communication**

The communication of the devices is supported by the communication layer. The protocols that can be used for connection between the devices and the cloud are numerous.

The three most reknown protocols are:

- HTTP/HTTPS
- MQTT 3.1/3.1.1 (Message Queuing Telemetry Transport) [19]
- Constrained application protocol (CoAP) [20].

Generally, IoT devices use the Internet protocol (IP) stack to connect to the WWW. This stack requires a lot of memory and processing power from the connected devices due to its complexity. In addition to connecting locally via non-IP networks, which use less energy, the IoT devices can also connect to the WWW through a smart gateway. Although quite common, some transmission channels like RFID, Bluetooth and NFC have a short range and their uses are therefore restricted to little personal area networks. There are various libraries that handle HTTP. Several small devices, like 8-bit controllers, can only partly implement the protocol because it is a simple text-based one; for example, they may only have sufficient code to POST or GET a resource. The more powerful 32-bit machines can make use of full HTTP client libraries that correctly implement the entire protocol. Many protocols have been tailored for Internet of Things use. The two most well-known are CoAP7 and MQTT6. MQTT was created in 1999 to address problems with SCADA and embedded systems. The OASIS MQTT Technical Committee8 is now standardizing the most recent version, 3.1.1, after several iterations. MQTT is a broker-based publish-subscribe messaging service. MQTT was created to operate over TCP. MQTT-SN, a related specification created for ZigBee-style networks, is also available (sensor networks). The IETF has developed a protocol called CoAP that aims to provide a RESTful application protocol based on HTTP semantics with a considerably lighter carbon footprint and a binary-based method as opposed to a text-based one. Instead of using a brokered strategy, CoAP uses a more conventional client–server approach. CoAP is made to be used instead of UDP. As the preferred device communication protocol for the architecture, we have chosen MQTT, with HTTP as a backup option [21].

IEEE 802.15.4, 6LoWPAN, Sigfox, RFID, NFC, LoraWAN, low-power WiFi, and other wireless network protocols are the main communication techniques utilized in the IoT sector. All these communication techniques are described below.

(a) **Near Field Communication (NFC)**

It is an extremely short-range wireless communication technique that enables interaction between mobile devices over just a few centimeters. If we bring two NFC-enabled devices near to each other, then they can start exchanging data by RFID technology. It operates near 14 MHz frequency band and provides two-way communication of data. By generating magnetic fields, both the devices can transfer data. The fact that close proximity between devices is necessary for safe transactions like payments is one advantage of this need. NFC-enabled technology is embedded inside all the smartphones in today's market [22].

(b) **Bluetooth Low Energy (BLE)**

Compared to rival methods, it uses less energy and has a somewhat shorter range. It is similar with classical Bluetooth technique [23]. It is composed of a controller and a host. The controller implements the physical and connection layers. Usually, the controller consists of a radio-equipped system on chip (SOC). The host includes the upper layers' functionality. The classic Bluetooth supports the streaming of data whereas BLE provides the transition of small data packets with 1 Mbps speed that results in a low duty cycle and less transmitting time. Devices in BLE can be either master or slave. One master connected with different slaves. Slaves should be extremely power-efficient in some circumstances. So, in order to conserve energy, slaves are set to sleep by default and awaken sometimes in order to receive data packets from the master. The IP-based communication is also permitted by the BLE protocol stack. BLE uses a lot less energy than competitive protocols like ZigBee. ZigBee is 2.5 times less energy-efficient compare to BLE.

(c) **IP-based WSN**

A wireless sensor network (WSN), which is made up of multiple sensor nodes connected by WiFi technology, allows for the monitoring of many sensor nodes spread across a wide region (unlike RFID, Bluetooth, NFC, etc.). They gather environmental data and transmit it to gateway devices, which then transmit it via the Internet to the cloud. In a WSN, nodes may communicate directly or via many hops. The gateway nodes have enough power and processing capability compared to the limited nature of the sensor nodes. A WSN frequently employs the star, mesh, and hybrid network topologies. The IEEE 802.15.4 standard serves as the foundation for the majority of WSN communication.

(d) **IPv6 over low-power wireless personal area networks (6LoWPAN)**

Because of its scalability, large addressing space, and reliability, IPv6 is regarded as the finest protocol for interaction in the space of IoT. IPv6 can be used over 802.15.4 wireless networks according to a specification. It permits IPv6-based communication over the IEEE 802.15.4 protocol. Between the transport layer and the link layer of 802.15.4, this standard specifies an adaption layer. All Internet protocol-based devices can interact with 6LoWPAN devices. 6LoWPAN networks use a gateway to access the Internet (WiFi or Ethernet). The 802.15.4 standard has 127-byte MTU which is too small to accommodate IPv6 headers. Then, the adaption layer optimizes the packets by compressing and fragmenting them to carry just the crucial data. It can perform optimization into three ways,

- To reduce IPv6 overhead, header compression of IPv6 packets has been determined by 6loWPAN. Several of the fields have been eliminated because of the ability to swap across packets.
- Fragmentation results breaking up the IPV6 packets into smaller units.
- Mesh under routing is also allowed by 6LoWPAN and is carried out at the link layer utilizing link-level short addresses as opposed to the network layer. Within a 6LoWPAN network, this functionality can be used for communication [24].

(e) **WiFi with Small Power**

Recently, the WiFi Alliance created "WiFi HaLow", a protocol based on the IEEE 802.11ah standard [25]. It has a broader coverage and uses less power than a typical WiFi gadget. This is why apps involving the Internet of Things can use this protocol. WiFi HaLow has almost double the range of regular WiFi. The supporting devices also allow IP connectivity suitable for IoT uses.

(f) **ZigBee**

For a wide variety of IoT applications, such as smart homes, remote controls, and medical systems, ZigBee smart energy (IEEE 802.15.4) is created. It supports several different network topologies, including as peer-to-peer, mesh, cluster-tree, and star. The ZigBee standard specifies the ZigBee and ZigBee Pro stack profiles. These stack profiles work with various applications and provide full mesh networking, enabling implementations with little memory and processing power. More capabilities are available with ZigBee Pro, such as security via symmetric key exchange, flexibility via stochastic address assignment, and improved performance via effective many-to-one routing techniques. The communication range of ZigBee devices is just 10–100 m. The ZigBee standard also includes information for the network and application levels. The network layer in this case supports multihop routing, unlike BLE [26]. A ZigBee network consists of three different kinds of devices: coordinator, router, and end devices.

- Every network must always have one of these since **coordinator** is the component that coordinates and builds the network. Other devices (routers or end devices) can connect to the network after this device establishes it. Its duties include choosing the frequency bands and giving devices their PAN IDs, or personal area network identification. Network devices communicate with one another using the PAN ID. Data routing over mesh networks may be assisted by the coordinator. It must have a constant power source, be operational at all times, and be capable of supporting child devices.
- **Router** must first connect to the network before it can permit other Routers and End devices to do the same. It must have a constant power source, be operational at all times, and be capable of supporting child devices.
- **End devices** have no connections to additional network hardware. They often run-on batteries and have a "sleep" mode for energy conservation.

Application objects (APO), application sublayer (APS), and a ZigBee device object make up the application layer (ZDO). The network nodes are covered in APOs which are software components that operate an underlying device hardware. The apparatus and the ZDO offer services for network administration which the APOs then employ. Transfer of data services is supplied to the APOs via application sublayer and ZDO. It is in charge of ensuring safe interactions between the objects that is utilized to develop a big scale kind of applications. Some common standards use ZigBee as a base. They are as follows,

- **ZigBee home automation** (ZHA) is a global standard for home automation that regulates things like lighting, temperature, energy use, security, and accident avoidance.
- Home area network electrical distribution companies and service providers can control energy use thanks to the **ZigBee smart energy** (ZSE) international standard. Suppliers and customers can communicate with one another using ZSE, giving both parties access to smart communications.
- An international standard called **ZigBee light link** (ZLL) enables consumer lighting components and other elements to communicate with one another, providing consumers with wireless access to these components. Customers can manage energy use and make their homes "greener" by controlling the lights in their homes.

(g) **Combination of WSN and RFID**

RFID can be used to identify object with little power and inexpensive way. Wireless sensor network (WSN) increases the usefulness of RFID through enhancing sensing capabilities, intelligence and multihop communication. There are three ways of combination possible between WSN and RFID [27].

(1) Sensor tags are RFID tags with built-in sensors. These sensor tags gather information about their surroundings, which the RFID reader may subsequently read. In these circumstances, straightforward RFID protocols with a single hop of communication are employed. In accordance with the power

requirements of sensor tags, which were previously discussed, RFID are two types, active and passive.

(2) Sensor tags are only capable of a single hop of interaction. The sensor tag has a wireless transceiver, a small amount of flash memory, and computing capability to increase its functionality. This allows it to start conversations wirelessly. In this way, wireless mesh network can be built. Sensor tags in these networks can interact with one another across a great distance. We can decrease the net quantity of data conveyed and boost the WSN's power efficiency by adding more processing power to each node.

(3) The range of RFID tag scanners is also increased with this kind of connection. The readers have wireless transceivers and microcontrollers so they may connect with one another. As a result, tag data can reach readers that are outside of a tag's range. It makes use of wireless sensor network devices' multihop communication.

(h) **DASH7**

DASH7 is a wireless active RFID communication protocol that works in the Industrial Scientific Medical (ISM) band, which is a globally accessible band, and is appropriate for IoT applications. In comparison to conventional ZigBee [26], it is primarily intended for scalable, long-range outdoor coverage with a higher data throughput. It is a cheap solution that works with IPv6 addresses and encryption. It is intended for burst, compact, asynchronous, and transitive traffic and provides a master/slave architecture. The filtering process for the incoming packets in DASH7 is very strict. The unique identity for DASH7 is the EUI-64 ID, and the dynamic network identifier is a 16-bit address chosen by the network administrator. Widespread use of RFID technology in WSN includes NFC and Dash7 (wireless sensor networking). Most smartphones will incorporate these technologies to offer users a variety of capabilities. Premises access, contactless payments, cutting-edge location services, home automation, tickets, and more fall under this category. Dash7 will therefore be a component of the Internet of Things (IoT), which collects sensor data and uses it to control social network applications. Pull and push are the two communication models that DASH7 allows. DASH7 consists of four device classes, blinker, endpoint, subcontroller, and gateway. Each device in the Dash7 network is compatible with at least one of the aforementioned device classes. Foreground frames and background frames are the different sorts of frames that DASH7 specifies. Regular messages that include data or data requests make up the foreground frames. Besides that, background frames are brief broadcast messages.

(i) **Low-Power Wide Area Networks (LPWAN)**

Low bitrate but long-range communication for IoT applications include the LPWAN group of protocols. The streamlined, lightweight protocols of LPWAN diminish hardware design complexity and equipment costs. Because of its extended range and star topology, less expensive infrastructure is needed, and using licensed or license-free bands lowers network expenses. Some impressive technologies used in this area are discussed below.

- **Narrow band IoT** is a technology designed for several energy-constrained devices. Therefore, it is required to lower the bit rate. However, both the GSM and LTE spectrum are suitable for this protocol. The downlink varies in speed from 40 kbps to 10 Mbps.
- Another protocol that makes advantage of narrow band (10 MHz) communication is **Sigfox**. It transmits its data via open wireless spectrum (the ISM band). Sigfox concentrates on using extremely long waves rather than 4G networks. As a result, the range can be extended to 1000 km.
- Operating in the sub-1 GHz range is the **Neul** protocol. In order to build less expensive, small power networks with extremely high scalability, it makes use of small portions of the TV whitespace spectrum. It communicates via the weightless protocol and has a 10 km range.
- LoRaWAN is a low-power protocol that is intended for wide area network operations. It is applicable in an urban or suburban setting, and its data rates range from 0.3 kbps to 50 kbps. It is one of the state-of-the-arts long-range Internet of Things protocols. As a result, it contains capabilities that support many tenants, allow for multiple applications, and include a variety of network zones.

(j) **Lightweight Application Layer Protocols**

Verbose web-based and data model protocols may be too heavy for IoT applications when constrained networks and/or a large-scale deployment of constrained nodes are taken into account. The IoT sector is developing new lightweight protocols to solve this issue and make them more appropriate for networks and nodes with a lot of constrained nodes. The two most widely used protocols are MQTT and CoAP. Their place in a typical IoT protocol stack is highlighted in Fig. 3.5.

iv. **Middleware**

When discussing how an IoT platform controls all communication between the hardware and the application layers and links remote devices to user applications (or other devices), it is frequently referred to as middleware [28]. It is also

Fig. 3.5 Protocol stack of lightweight application layer protocols

CoAP	MQTT
UDP	TCP
IPV6	
6LoWPAN	
802.15.4 MAC	
802.15.4 PHY	

referred to as a cloud enablement platform or an Internet of Things enablement platform to highlight its key commercial benefit, which is the addition of cloud-based software and services to everyday devices. Lastly, it changes the emphasis to becoming a crucial tool for IoT developers under the banner of the IoT application enablement platform. Data gathering from the devices using various protocols and network topologies, remote device configuration and control, device administration, and over-the-air firmware updates were among its main functions. IoT middleware is anticipated to facilitate integration with practically any connected device and blend in with third-party apps used by the device in order to be used in heterogeneous IoT ecosystems in real-world settings. A single IoT platform can easily handle any type of connected device due to its independence from underlying hardware and overarching software (shown in Fig. 3.6). In addition, current IoT platforms add a number of useful capabilities to the hardware and application levels. They offer parts for cloud-based deployment, on-device data processing, and frontend and analytics. A few of them can manage the complete end-to-end installation of an IoT system.

Some well-known IoT middleware systems are described below.

a. The EU promotes **FiWare**, a very well-liked IoT middleware system. It was created with logistics, shop floor analytics, and smart cities in mind. Numerous FiWare developers have contributed a sizable body of code, reusable modules, and APIs to the project. Any application developer can create their own IoT application using a portion of these components. Context information is the term FiWare uses to describe the data that sensors gather. In order to capture the context from many circumstances, it defines generic REST APIs. A specialized service called a context broker receives all of the context information. FiWare offers APIs that allow you to query and save the context. Additionally, any application can sign up to be a context

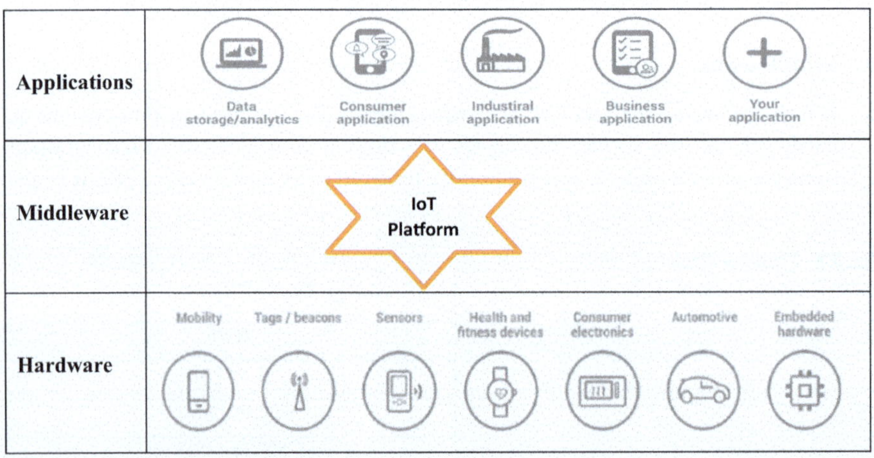

Fig. 3.6 IoT platform as middleware

consumer and make information requests to the context broker. Additionally, FiWare describes a collection of SNMP APIs through which we may setup and handle the actions of IoT devices.

b. Another well-known open-source project is **OpenIoT**. It is made up of 7 separate parts. We have a physical plane as the most fundamental level. It gathers data from IoT equipment and performs preparation on that data. This includes many APIs that allow it to communicate with various kinds of physical nodes and obtain data from them.

3.3 Industry Applications of IoT

IoT has a huge number of applications ranging from smart health to smart automobiles.

- IoT plays a vital role in urban agriculture. The plant environment can be monitored, managed, and customized using sensors and actuators to meet the requirements of the particular specimen. On demand, water can be supplied via both remote feeds and rainwater collecting. Fertilizer supply can also be done automatically by remotely operated drones. Blinds that are required to shield and protect as well as produce greenhouse microclimates can be automatically regulated, and the weather and light can be controlled [29].
- IoT has a valuable impact on the mining industry with the help of building robotic equipment. Mine sites will be connected, heavy equipment will be remotely operated and monitored, and air and gas monitoring of shafts will be done. Sensors helps to find the location of the machines inside the mine.
- IoT also shows its applications in supply food chain management. There will be a connection between the logistics chain to track environmental conditions as product is carried, the tracking of farming conditions for the health of plants and animals, and the handling of food by merchants.
- Landslides and other natural calamities can be predicted, and preparedness measures can be taken, thanks to the sensors' autonomous synchronization and simulation.
- The IoT can assist in the design of smart homes, including energy usage control, appliance interactivity, emergency detection and easy item searching, home security, etc.
- The Internet of Things (IoT) can also be used in the medical field to save lives or enhance quality of life, such as by tracking health metrics, activities, promoting independence, and monitoring medication intake, among other things [5, 30].
- The IoT can assist in the creation of smart cities by, for example, monitoring air quality, identifying escape routes, illuminating the city effectively, watering gardens, and other functions [31].

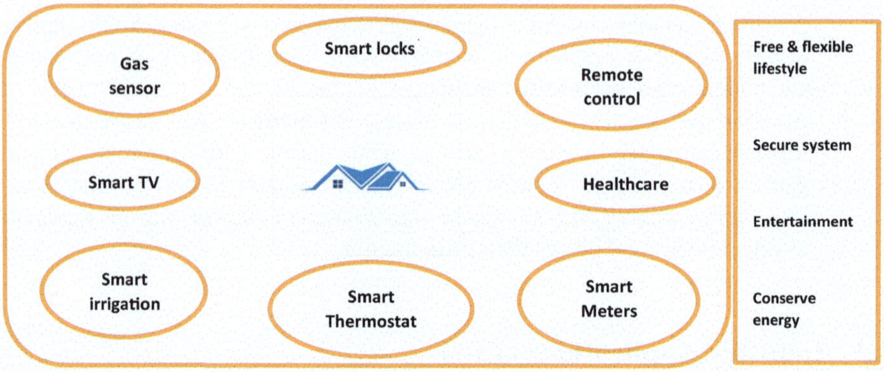

Fig. 3.7 A sample smart home system [32]

- Bidirectional data transfers made possible by IoT will improve how self-driving cars communicate with the transportation network around them and give passengers access to vital information. To function to their full capacity, self-driving cars require always-on, dependable connectivity as well as information from other transportation-related sensors. The term "connected roads" refers to driving and driverless vehicles that are completely integrated with the local transportation network. Google's self-driving car is an example of it.

There are a set of challenges facing by IoT.

- IoT technology establishes a network of interconnected devices, increasing security. Despite adequate security protections, the system may provide limited authentication control during this process.
- Without the user's voluntary engagement, the use of IoT discloses a significant amount of personal data in great detail. This raises numerous privacy concerns.
- The IoT system's flexibility is a source of worry. Since there are so many different systems engaged in the process, the key concern is connecting with another system.
- The IoT system's design is also highly intricate. Furthermore, it is difficult to deploy and maintain. IoT has its own guidelines and standards, which must be followed. The task of compliance is difficult due to its complexity, though (Fig. 3.7).

3.4 Big Data Analytics via IoT with Cloud Service

IoT, big data, and cloud computing's interaction offers plenty of opportunities for businesses to achieve rapid growth. Simply defined, IoT is the data source, big data is the platform for data analytics, and cloud computing is the place for preservation, scaling, and access speed. IoT devices frequently gather enormous amounts of data. It is essential to thoroughly analyze all of this data. In order to analyze IoT data, numerous big data methods are employed. Additionally, it is likely that some of the data gathered may be lacking due to the network's frailty. This must be considered, and data must be extrapolated using powerful machine learning methods. In order to deliver highly scalable map reduce analytics on the data arriving from the devices, big data analytics is a cloud scalable platform that supports technologies like Apache Hadoop [33, 34].

A deployment of the "Internet of Things" must include the cloud. The majority of the data collected from sensors is evaluated and kept in one main cloud. IoT middleware is required to enable customers to utilize the cloud to improve their understanding of the data gathered by the sensors and to function seamlessly on a variety of clouds. In the end, big data and Internet of Things work together to achieve shared objectives by turning data into useful information for enterprises. For instance, a larger picture of a company's total performance over time can be obtained by combining real-time IoT information with long-term big data analytics. Specific industries will profit as these technologies advance. The use of IoT and big data will be advantageous for sectors including the transportation, shipping, and automobile industries. The insights and knowledge gleaned from the technologies will enhance the analytics process, making it quicker and simpler, improving efficiency, saving money, and encouraging well-informed forecasts. Typical relational databases (RDBMS) are not suitable for handling IoT data [34, 35]. RDBMS face two issues while handling IoT data,

- **Scaling Issue**: Relational databases can get very big very quickly since most IoT networks have a lot of smart objects that send data all the time. This may lead to performance problems that can be expensive to fix and frequently call for additional hardware and architectural changes.
- **Data Volatility Issue**: It is crucial that the schema be created appropriately from the start when using relational databases. Later changes could cause the database to run slowly or not at all. Revisions to the schema must be maintained to a minimal due to its rigidity. However, IoT data is unstable in that the data model may alter and develop over time. In order to allow for daily or even hourly data model updates, a dynamic schema is frequently needed.

To handle and overcome all these challenges NoSQL database is used. A database without SQL is known as a NoSQL database. It is not organized in a relational database's customary tabular format. NoSQL databases provide for a flexible data model and do not enforce a rigid format. Additionally, these databases are far more scalable by nature. Besides that, IoT brings another challenge of handling large size streaming data at the network level. It is difficult to analyze and respond to streaming data, which is generated as smart objects transmit data, because it is typically quite large in volume and has little value unless it can be done so. The ability to analyze streaming data in real-time enables you to look for trends or anomalies that can point to a problem or a circumstance that requires an instant response. You need to be able to filter and evaluate the data as it is happening, as near to the edge as you can, in order to have a chance of changing how the situation turns out. Real-time streaming data analysis is rapidly expanding as a market. Streaming analytics are a feature of major cloud analytics providers like Google, Amazon, Microsoft, and IBM, and numerous additional programs can be utilized internally. Tools for network analytics like Flexible NetFlow and IPFIX enable the detection of anomalous patterns or other issues in the transmission of IoT data via a network. The entire processes of data acquisition, data storage, and computing are described in brief.

3.4.1 Data Acquisition, Preprocessing, and Storage

There various types of data generated from IoT devices (like, RFID, or ATM devices), and finally, it transfers to cloud storage through Internet gateways. Data acquisition refers to gathering information from M2M or IoT devices, following interactions with a data acquisition system, the data communicates (application). To obtain the required data, the program connects and communicates with a variety of devices. The gadgets transmit data when requested or at predetermined intervals. Using the network, transport, and security layers, devices transmit their data.

Data obtained from the devices does not necessarily imply that they are accurate, significant, or consistent. Data consistency refers to data that is within the predicted range, consistent with a pattern, or not corrupted during transmission. Data needs to be validated, therefore. Software for data validation performs validation checks on the collected data. Software for validation uses logic, rules, and semantic annotations. Applications and services rely on accurate data. Only then can the analytics, forecasts, recommendations, diagnoses, and judgments be considered valid. Big data is another category for data that is categorized from many different devices and sources. Big data is data that is kept in databases on a server, in a data warehouse, or on the cloud. Besides big data, there are three categories of data that are required to be stored.

a. Data that must be stored alone because it will need to be processed, referred to, or audited regularly in the future.
b. Data that only needs to be processed once, and the results are used later using analytics; the data as well as the outcomes of processing and analytics is saved. Fast visualization and report production without reprocessing are benefits of this case. Additionally, the data is accessible for future audits or references.
c. Streaming, real time, or online data must be processed, and the analysis's findings must be stored.

Big data is a fourth category for data that is categorized from many different devices and sources. Big data is data that is kept in databases on a server, in a data warehouse, or on the cloud. Think about products having RFID tags. Applications for tracking or inventory control require both the locations and IDs of the commodities when they travel from one location to another. Storage in the form of a spatial database that is optimized for storing information is referred to as spatial storage. Let us say a city needs a computerized map to find parking spots. Data that depicts things specified in a geometric space is referred to as spatial data. Figure 3.8 shows the relationship between IoT and big data analytics [30].

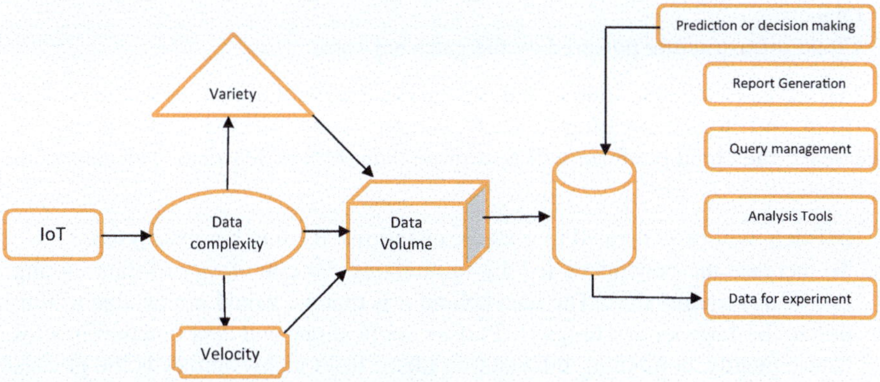

Fig. 3.8 Relationship between IoT and big data analytics

3.4.2 Computing in Cloud Framework for IoT

A modern approach to data acquisition, storage, and computation is the cloud. A centrally located server with accessible computing resources is known as the cloud. IoT-generated big data packages can easily flow via the Internet using cloud computing. For data collecting, storage, computation, and services, use the cloud computing paradigm. Anything as a service, a software architecture paradigm, defines cloud computing service concepts. We can highlight cloud-based IoT services like Xively and Nimbits and also explain platforms including TCS Connected Universe, IOx and Fog, IBM IoT Foundation, AWS IoT, and Cisco IoT [36].

In conjunction with the Internet of Things, cloud computing helps to increase the effectiveness of routine jobs. The goal of cloud computing is to give data a way to get where it needs to go, yet the Internet of Things generates a ton of information. Amazon Web Services lists the following four advantages of cloud computing:

- No need to estimate infrastructure capacity requirements.
- Savings increase with the size of the scale because you only need to pay for the resources you really use.
- Platforms can be set up all over the world in a short amount of time.
- Flexibility and promptness in supplying developers with resources.

The new idea of an Internet network has developed because to the Internet of Things. This permits communication between a numbers of items, including

Smart gadgets.
Mobile gadgets.
Various sensors.

All architectural components can effectively communicate with one another thanks to the Internet of Things' architecture. The components could

Objects.
Cloud infrastructure powered by the gates network.

IoT serves as the data source, and big data as a technology serves as the analytic platform, and cloud computing functions as their shared workspace. Within the next five years, the cloud platform will host more than 90% of IoT data. The causes are as follows:

1. IoT data will be generated in massive quantities, feeding big data systems.
2. To increase the benefits of IoT by minimizing the complexity of data mixing, one of its requirements. The idea behind it is that we would not be able to fully utilize the Internet of Things (IoT) if its applications and data function in silos. Consequently, combining information (data) from many sources is the greatest strategy to gain superior insights and make judgments.

We therefore see a clear necessity for embracing cloud-based technologies for both IoT and big data in light of the two considerations indicated above.

Therefore, cloud computing's function in the Internet of Things is to pool resources to store IoT data and make it accessible when needed. It is crucial to remember that using cloud computing makes it simple to send huge data packets produced by the IoT across the Internet. According to IoT, the Internet business will have $20 trillion in annual sales in the next ten years, creating opportunities for greater research and development in the big data and IoT fields [33].

Benefits of Trios

- In order to accommodate the needs of hosting and analyzing big data, cloud-based systems can be scaled both vertically and horizontally. For instance, you can add more applications to a server to improve its capacity, or you can increase your physical resources as needed. The development of big data and data analytics is made possible by the cloud.
- Large amounts of data can be stored using both big data and cloud data, which also offers scalable processing and improved real-time data analysis. Big data, IoT, and the cloud do not require any physical infrastructure to get started, which lowers expenses and frees you up to concentrate on the enhanced analytical power rather than worrying about upkeep and care.
- The cloud offers the data's route of transit because IoT and big data generate a lot of data.
- When using the cloud, you may simply and remotely access big data from anywhere in the globe to continue performing tasks on devices, facilitating improved cooperation.
- The requirement for intelligent devices to deliver data to servers for processing rather than to central servers arises from the demand that a large number of devices can put on an Internet connection. You may now access data from many network locations, be on the "edge" of processes, respond to outages faster, and anticipate when issues can happen. Additionally, combining the cloud with IoT helps to

improve security because regular updates can be given, and any infrastructure breaches can be quickly reported.

- By effectively storing and managing your big data and IoT with the cloud, which includes built-in management tools, processing power, and applications to monitor your resources, you can maintain business value.

By combining big data, IoT, and the cloud, you can successfully link and move data across devices in the most efficient and effective manner possible. In short, cloud computing is a facilitator. It makes IoT and big data hosting platforms as well as process and data analytics possible. The key advantage of combining cloud computing with IoT and big data is that it offers enterprises a scalable, dependable, and adaptable solution. IoT, big data, and cloud computing are interdependent in a synergistic way that enables your company to get actionable insights through analysis and performance reports [37, 38].

3.5 Conclusion

To sum up, the Internet of Things, big data, and cloud computing convergence open up a new frontier for decision support systems. Additionally, the IoT, big data, and cloud computing convergence may open up new possibilities and applications across all industries. Besides that, this will provide fantastic career opportunities for experts who are currently working on the specific technology. The Internet of Things and cloud computing will fundamentally alter human life, especially in terms of how information is managed. Depending on the deployment strategy, the cloud is the sole technology that can analyze, store, and access IoT data. Due to the nature of on-demand information, cloud computing is accessible whenever and wherever there is an Internet connection. Many businesses are becoming aware of the advantages of hybrid cloud adoption and the necessity to adopt it. For a very long time to come, cloud computing will keep creating new possibilities for IoT.

References

1. Vermesan O, Friess P, Guillemin P et al (2011) Internet of things strategic research roadmap. Internet Things: Glob Technol Soc Trends 1:9–52
2. Said O, Masud M (2013) Towards internet of things: survey and future vision. Int J Comput Netw 5(1):1–17
3. Atzori L, Iera A, Morabito G (2010) The internet of things: a survey. Comput Netw 54(15):2787–2805
4. Wójcicki K, Biegańska M, Paliwoda B, Górna J (2022) Internet of Things in industry: research profiling, application, challenges and opportunities—a review. Energies 15(5):1806
5. Islam M, Nooruddin S, Karray F, Muhammad G (2022) Internet of things device capabilities, architectures, protocols, and smart applications in healthcare domain: a review. arXiv preprint arXiv:2204.05921

6. Weyrich M, Ebert C (2016) Reference architectures for the internet of things. IEEE Softw 33(1):112–116
7. Gubbi J, Buyya R, Marusic S, Palaniswami M (2013) Internet of Things (IoT): a vision, architectural elements, and future directions. Future Gener Comput Syst 29(7):1645–1660
8. Aazam M, Huh E-N (2014) Fog computing and smart gateway based communication for cloud of things. In: Proceedings of the 2nd IEEE international conference on future internet of things and cloud (FiCloud '14), Barcelona, Spain, Aug 2014, pp 464–470
9. Lv M, Chen L, Chen G, Zhang D (2014) Detecting traffic congestions using cell phone accelerometers. In: Proceedings of the international joint conference on pervasive and ubiquitous computing (UbiComp '14), Seattle, Wash, USA, Sept 2014, pp 107–110
10. Anjum A, Ilyas MU (2013) Activity recognition using smartphone sensors. In: Proceedings of the IEEE 10th consumer communications and networking conference (CCNC '13), Las Vegas, Nevada, USA, Jan 2013, pp 914–919
11. Bui N, Zorzi M (2011) Health care applications: a solution based on the internet of things. In: Proceedings of the 4th international symposium on applied sciences in biomedical and communication technologies (ISABEL '11), ACM, Barcelona, Spain, Oct 2011
12. Pantelopoulos A, Bourbakis NG (2010) A survey on wearable sensor-based systems for health monitoring and prognosis. IEEE Trans Syst Man Cybern Part C Appl Rev 40(1):1–12
13. Gupta N, Khosravy M, Patel N, Dey N, Gupta S, Darbari H, Crespo RG (2020) Economic data analytic AI technique on IoT edge devices for health monitoring of agriculture machines. Appl Intell 50(11):3990–4016
14. Frank K, Robertson P, Gross M, Wiesner K (2013) Sensor based identification of human stress levels. In: Proceedings of the IEEE international conference on pervasive computing and communications workshops (PerCom Workshops '13), San Diego, CA, USA, Mar 2013, pp 127–132
15. Tallapragada VV, Kullayamma I, Kumar GV, Venkatanaresh M (2022) Significance of Internet of Things (IoT) in health care with trending smart application. In: Smart systems: innovations in computing. Springer, Singapore, pp 237–245
16. Sekhar PK, Brosha EL, Mukundan R, Garzon FH (2010) Chemical sensors for environmental monitoring and homeland security. Electrochem Soc Interface 19(4):35–40
17. Welbourne E, Battle L, Cole G et al (2009) Building the internet of things using RFID: the RFID ecosystem experience. IEEE Internet Comput 13(3):48–55
18. Mukherjee A, Dey N, De D (2020) EdgeDrone: QoS aware MQTT middleware for mobile edge computing in opportunistic Internet of Drone Things. Comput Commun 152:93–108
19. Stanford-Clark A, Linh Truon H (2008) MQTT for sensor networks (MQTT-S) protocol specification. International Business Machines Corporation Version 1
20. Shelby Z, Hartke K, Bormann C (2014) The constrained application protocol (CoAP). Tech. Rep., IETF
21. Vimal S, Khari M, Dey N, Crespo RG, Robinson YH (2020) Enhanced resource allocation in mobile edge computing using reinforcement learning based MOACO algorithm for IIOT. Comput Commun 151:355–364
22. Coskun V, Ozdenizci B, Ok K (2013) A survey on near field communication (NFC) technology. Wirel Pers Commun 71(3):2259–2294
23. Chang K-H (2014) Bluetooth: a viable solution for IoT? [Industry Perspectives]. IEEE Wirel Commun 21(6):6–7
24. Culler D, Chakrabarti S (2009) 6lowpan: incorporating IEEE 802.15.4 into the IP architecture, IPSO Alliance. White Paper
25. Shanmuga Sundaram B (2016) A quantitative analysis of 802.11Ah wireless standard. Int J Latest Res Eng Technol 2
26. Baronti P, Pillai P, Chook VWC, Chessa S, Gotta A, Hu YF (2007) Wireless sensor networks: a survey on the state of the art and the 802.15.4 and ZigBee standards. Comput Commun 30(7):1655–1695
27. Mitrokotsa A, Douligeris C (2009) Integrated RFID and sensor networks: architectures and applications. In: RFID and sensor networks: architectures, protocols, security and integrations. Auerbach Publications, pp 511–535

28. Razzaque MA, Milojevic-Jevric M, Palade A, Cla S (2016) Middleware for internet of things: a survey. IEEE Internet Things J 3(1):70–95
29. Sinha BB, Dhanalakshmi R (2022) Recent advancements and challenges of Internet of Things in smart agriculture: a survey. Future Gener Comput Syst 126:169–184
30. Sriramalakshmi P, Vondivillu ST, Govind ASK (2022) Role of IoT, AI, and big data analytics in healthcare industry. In: Transformation in healthcare with emerging technologies. Chapman and Hall/CRC, pp 19–43
31. Fantin Irudaya Raj E, Appadurai M (2022) Internet of Things-based smart transportation system for smart cities. In: Intelligent systems for social good. Springer, Singapore, pp 39–50
32. Yu M, Rhuma A, Naqvi SM, Wang L, Chambers J (2012) A posture recognition-based fall detection system for monitoring an elderly person in a smart home environment. IEEE Trans Inf Technol Biomed 16(6):1274–1286
33. Dey N, Hassanien AE, Bhatt C, Ashour A, Satapathy SC (eds) (2018) Internet of things and big data analytics toward next-generation intelligence, vol 35. Springer, Berlin
34. Sarowar MG, Kamal MS, Dey N (2019) Internet of Things and its impacts in computing intelligence: a comprehensive review—IoT application for big data. Big data analytics for smart and connected cities, pp 103–136
35. Bhatt C, Dey N, Ashour AS (eds) (2017) Internet of things and big data technologies for next generation healthcare
36. Dittakavi S, Bhamidipati G, Neelam V (2018) Big data analytics via IoT with cloud service. In: Big data analytics. Springer, Singapore, pp 319–328
37. ur Rehman MH, Yaqoob I, Salah K, Imran M, Jayaraman PP, Perera C (2019) The role of big data analytics in industrial Internet of Things. Future Gener Comput Syst 99:247–259
38. Lv Z, Qiao L, Verma S (2021) AI-enabled IoT-edge data analytics for connected living. ACM Trans Internet Technol 21(4):1–20

Chapter 4
Virtual and Augmented Reality with Embedded Systems

4.1 Introduction

Virtual reality (VR) is a computer-generated world that is presented to the viewer in such a way that it appears and feels genuine. The user must put on specialized gloves, ears, and goggles in order to enjoy a virtual reality world. These devices take input from the computer system. The user's actions are continuously seen and analyzed by the computer, which also modifies the data given to the devices they are wearing. For instance, the goggles track head motions and react by transmitting a new video input, giving the user the impression that they are in a real environment. The environment being simulated may resemble the real world, as in pilot or military training simulations, or it may be very different, as in the exotic worlds and animals seen in virtual reality games. The objective of augmented reality (AR), in contrast, is to give information and meaning to a genuine object or location. In contrast to virtual reality, augmented reality does not attempt to create a computer-simulated environment. Instead, it builds on the real object or place that the user is observing and adds context information to enhance understanding of the object. A user's view of an object or a place can be enhanced by adding data from augmented reality, such as an audio, geographical information, historical context, or other information. The majority of existing augmented reality (AR) apps are created using general-purpose processors, mainly using softwares. However, it does not guarantee best performance due to its high intensive requirements and power consumption. On the other hand, the central component of every digital gadget is the embedded system. For the design and development of AR to simplify difficult concepts, a solid understanding of the embedded system is required.

S. Chakraborty and L. Dey, *Computing for Data Analysis: Theory and Practices*,
Data-Intensive Research, https://doi.org/10.1007/978-981-19-8004-6_4

4.2 Types of Augmented Reality Systems

The best aspect about augmented reality is how easily it can be used by regular people [1]. There are 2 types of AR systems: Marker-based augmented reality and markerless augmented reality (shown in Fig. 4.1).

Marker-based AR: Maker-based AR functions by scanning a marker, which causes an augmented experience to show on the device (which might be an object, text, video, or animation). To scan markers from a device utilizing its camera feed, it typically requires software in the form of an app. Games like Pokemon Go and Instagram, Facebook, and Snapchat filters use the marker-based augmented reality.

Markerless AR: In contrast to marker-based AR, which requires a marker to be scanned, markerless AR does not. It can be started with the web or mobile AR experience. The digital element is then placed on a flat surface, such the ground or a wall, once the real environment around the object has been scanned. Location-based AR, projection-based AR, superimposition-based AR, and outlining-based AR are the four subcategories of markerless AR.

Location-based AR: Location-based GPS, accelerometer, digital compass, and other technologies are used by augmented reality to pinpoint a device's precise location. The necessary sensors are present in the majority of contemporary mobile devices, making this potent technology accessible to all mobile users.

Projection-based AR: Projection-based AR is defined as a video projection approach that can add to and reinforce visual data by projecting images onto the surfaces of 3D objects or space. It is simple to build graphical representations that conventional lighting techniques cannot express using projection-based AR.

Superimposition AR: In superimposition AR, an augmented view of the same object replaces all or part of the original view of the object. Because an app cannot replace an original object with an augmented one if it cannot identify the original thing, object recognition is essential in this sort of AR.

Outlining AR: To aid in specific circumstances, specialized cameras are created for human eyes to execute outlining of the specified objects like limits and lines. Object recognition is used in outlining AR to improve knowledge of the existing world.

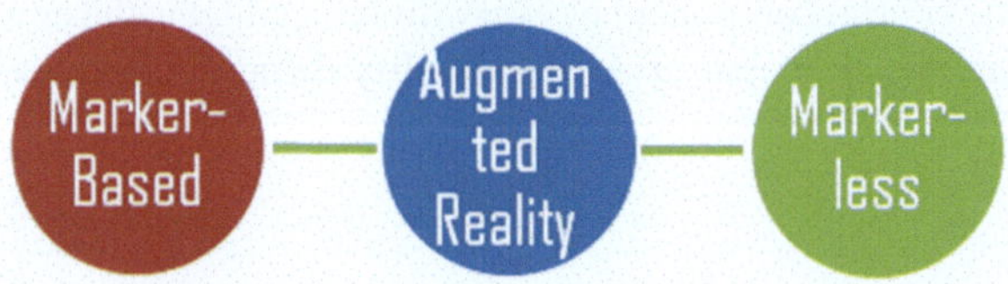

Fig. 4.1 Augmented reality types

4.3 Overview of Augmented Reality System Organization

4.3.1 History of Augmented Reality

The original idea behind augmented reality (AR) was started in the year of 1950s to support imaginative storytelling in plays, novels, and movies. Ivan Sutherland, an American engineer, is frequently cited as one of the founding fathers of augmented reality. He created the Sketchpad, a piece of 3D modeling and visual simulation software, in the 1960s. The University of Salt Lake City in Utah (USA) introduced the "head-mounted three-dimensional display" in 1968, a set of spectacles that allowed users to view images in three dimensions (Fig. 4.2). These were the real forerunners of Google glass (Fig. 4.2). The EyeTap, a helmet that projects virtual information in front of the user's eye, was created by Steve Mann in 1980. This is the first operational version of an augmented reality headset. The EyeTap is still in use today, but it has been refined to be as unassuming and understated as a straightforward pair of glasses.

Although the term augmented reality (AR) has been used since the 1990s, it has only lately begun to be widely used. The creation of AR applications has exploded since the introduction of Google's ARCore and Apple's ARKit in September 2017 and October 2017, respectively. With just a regular camera and no specialist hardware, these platforms have made it simple for anybody with an iPhone or Android phone to start creating augmented reality experiences.

Snapchat democratized augmented reality for all users on social media, often without them even realizing it. In 2014, the app debuted "geofilters", or filters that would adjust based on your location. These were actually additional "frames" or items that users may put on their screen to show their subscribers where they were. The 2016-released mobile application put augmented reality at the center of its gameplay. In order to find Pokemon or to win battles against other players, the players roam around in their actual surroundings while using their phones. Due to the huge success brought by the use of augmented reality in mobile gaming, numerous additional AR-based video games followed, including The Walking Dead: Our World, Angry Birds AR and Harry Potter: Wizards Unit (Fig. 4.3).

Most modern augmented reality (AR) apps are created on platforms with general-purpose processors, with software taking care of the processing duties. However,

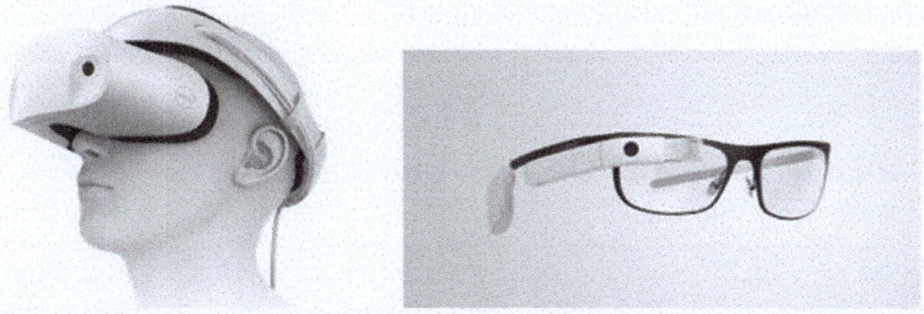

Fig. 4.2 Head-mounted display and Google glass

Fig. 4.3 AR game using handheld devices

due to the high-intensity requirements of the various processing activities involved in AR, software execution is not always the ideal choice. It always places constraints on frame rate and latency, endangering real-time operation, and increases size and power consumption, hindering mobility. These limitations make it harder for AR applications to spread. Today, embedded AR systems based on field-programmable gate array (FPGA) hardware and electronic system-level (ESL) technologies are employed to overcome these restrictions [2].

A computer hardware and software combination created for a particular purpose is an embedded system. Typical desktop or laptop hardware might prove to be too heavy, power-hungry for certain applications, where embedded systems being a self-contained, specialized computers offer an alternative. The mechanism used to determine the user's viewpoint in respect to the real environment is a crucial factor in any augmented reality application. There are different tracking methods used in embedded system; however, research makes use of a camera and computer vision algorithms to recognize specific markers in AR.

4.3.2 Embedded Systems Design Approaches

There has been very little research done on embedded systems for augmented reality applications, especially when it comes to embedded computer vision. Thankfully, a substantial body of research has been done on more generic embedded computer vision architectures. It is important to look at the embedded computer vision platforms because the tracking algorithm is one of the two main functions of a vision-based augmented reality system [3]. Depending on which of the numerous different CPU types is used, embedded computer vision platforms come in a variety of shapes and sizes [4]. A device can be customized to a specific purpose using specialized processing components like digital signal processors (DSPs) and field-programmable gate arrays (FPGAs).

The choice of an architecture will be determined by the limitations imposed by the application's design. Different architectures have different strengths and shortcomings. Higher-level algorithms containing branches and loops are better suited to DSPs, but low-level algorithms with a high degree of parallelism are a suitable fit for FPGAs. In order to allow high performance vision systems, for example, a DSP and FPGA combination could be used to make up for a device's weaknesses [5].

4.3.2.1 System Architecture

The capabilities of mixed reality, augmented reality, and VR are varied. All of these are utilized to build a completely virtual environment that enables people to experience an alternative environment. Computer vision is needed to accurately distinguish and categorize the surroundings so that AR can add the virtual world to them. Simultaneous localization and mapping (SLAM) is a computer vision technique that compares visual information between camera frames to map and track the surroundings. SLAM is frequently used in AR technologies. It is feasible to accomplish very accurate tracking when using the accelerometer and gyroscope data from a smartphone. The number of use cases that can be implemented can be considerably increased by adding machine learning algorithms on top of the fundamental AR technology.

A microcontroller-based system with software and a control system that can be utilized to carry out a certain task is referred to as an embedded system. The term "embedded system" has a broad definition that refers to a hardware system that includes software that is integrated into the hardware system. The embedded system may be a standalone unit or a component of a larger system. Software is used to distinguish the embedded system and can be used to control the embedded system. Because they are affordable and simple to use, embedded systems are commonly used in electronic products. The software is housed in the hardware system known as the embedded. The embedded system can be thought of as a microprocessor that can carry out a certain task.

There are primarily three categories of components in an embedded system:

- Hardware system.
- Application software.
- Real-time operating system (RTOS).

The system that holds the parts that can be utilized to construct an embedded system is known as the hardware system. To keep the total size of the embedded system modest, the size of the components is maintained to a minimum. The application that lives in the hardware system and can be utilized to carry out a certain task is known as the software system. One program can run in the system at a time, allowing for quick task processing and accurate results. The software used in embedded systems is unique to this kind of system and cannot be used as generic software for other computing devices due to the differences in hardware components that may result in setup issues. The real-time operating system is a kind of utility software that enables the processor to carry out operations sequentially and with little lag. A real-time monitoring system has the ability to directly control the whole operation of an embedded system [6].

Basic elements of embedded system architecture are defined in the following description:

- **Sensor**: A sensor is a tool used to transform physical quantities into electrical signals. The A-D convertor is utilized as a tool for converting sensor devices.
- An **A-D converter** is a device that transforms analogue signals into digital signals.
- **Processor**: A processor is a tool that can be used to process output and save the outcome to memory. The processor is the primary element of an embedded system. As input is received, processed, and output is generated using the processor.
- **D-A Convertor**: This is the category of equipment used to transform digital signals into analogue signals.
- **Actuator**: An actuator is a particular kind of device that contrasts the actual outcome with the outcome generated by the D-A convertor tool.

A. **Embedded System with Computer Vision Framework**

Depending on which of the numerous different CPU types is used, embedded computer vision platforms come in a variety of shapes and sizes. A device can be customized to a specific purpose using specialized processing components like digital signal processors (DSPs) and field-programmable gate arrays (FPGAs). The choice of an architecture will be determined by the limitations imposed by the application's design. Different architectures have different strengths and shortcomings. A developer can combine these processing components in many ways to create a finished product. The simplest method is to employ common PC hardware that is accessible in an embedded form-factor, as that described by the PC/104(1) standard. Auxiliary boards that add DSPs and FPGAs as well as other processing resources can be added to this base platform. The power these systems use is one of its flaws. An average power budget could be around 50 W. On the other hand, using such hardware eliminates the challenges of mimicking or porting to a different architecture, allowing an application to be developed on a desktop PC [7]. Smart cameras adopt a tightly integrated design strategy, in contrast to PC-based solutions. These kinds of gadgets have a processing component that is closely connected to the image sensor, creating a single unit that can carry out complex video analysis in real time. The "SmartCam" system's array of SIMD processing components and optional instruction-level-parallel processor was designed using this strategy. Less exotic processors, like a microcontroller or DSP, have been used in more contemporary systems [8].

Because these CPUs lack a memory-management unit (MMU), creating applications for them can be challenging. The MMU manages virtual memory, which is necessary for multitasking in standard operating systems. Instead, a non-standard OS must be used (such as uClinux2), or the application developer may decide to do without an OS altogether. Utilizing several processors in a single system is one approach for applications where no single processing device can deliver adequate computational capability. The multiprocessor system-on-chip (SoC), which combines the processors with components like embedded DRAM, flash memory, application-specific hardware accelerators, and RF components onto a single package, is the result of this. The potential openness of the software architecture of generic microprocessors and the effectiveness of synthesized hardware are benefits of this method.

B. **Embedded System with Mobile Devices**

The mobile communications sector has found extensive use for the SoC-type device. Modern cell phones come in a small, handy package that packs a lot of power and capabilities [9]. Although Samsung and Qualcomm also provide comparable chips, Texas Instruments' OMAP3430 serves as the N900's processor. Each chip has a unique feature set that changes depending on the manufacturer and the model, but all such devices have the following traits:

- One or more cores of an ARM processor.
- Extra processing components that are optional, like a DSP or graphics accelerator.
- Circuitry for memory control that interfaces with external RAM.
- A large collection of standard peripheral interfaces.
- A direct memory access (DMA) controller for swift data transmission.

Dual ARM Cortex-A9 CPUs, hardware video and graphics accelerators, and controllers for a range of peripherals are all included in this device. SDRAM is installed as an extra chip in a package-on-package (PoP) layout on top of the primary CPU. It is intriguing to observe how closely the feature set of the N900 matches the functional blocks in the diagram. For instance, the Nokia device makes advantage of the built-in LCD, camera, and memory card connections (all of which are present in the OMAP3430). The capacity of these CPUs to run a complete multitasking operating system is a crucial feature. For instance, the N900 uses the Linux-based Maemo OS, while many more utilize the Google-created Android OS, which is also based on Linux. A virtual reality headset called the Samsung Gear VR was created by Samsung Electronics in partnership with Oculus VR and is produced by Samsung. The Galaxy S7, S7 edge, Note5, S6 edge+, S6 and S6 edge are all compatible with the most recent version of the Samsung Gear VR. On Samsung VR, you can submit and share your own 360-degree videos in addition to accessing thousands of other premium 360-degree experiences. When in use, a compatible Samsung Galaxy device serves as the headset's display and processor, and the Gear VR unit itself serves as the controller, housing the field of view and an inertial measurement unit (IMU) specifically designed for rotational tracking that connects to the smartphone via USB-C or micro-USB. Along with a touchpad and back button on the side, the Gear VR headset also has a proximity sensor that detects when the headgear is on [10, 11].

C. **Embedded System with Augmented Reality Platform**

In addition to creating the virtual object overlay, the first embedded AR system utilized multiple TMS320C40 DSPs to implement object recognition and feature point detection algorithms. This system ran at 30 frames per second in real time (the framerate of the camera used) [12]. An augmented reality system tracked images using the FPGA technique. This FPGA-based system was created as a standalone unit, unlike the other FPGA-based systems previously discussed, which needed a supporting computer. An image sensor that is controlled via the I2C image sensor protocol provides a video stream. The video decoder interface consumes image data, whereas the video encoder interface generates it. Two parallel data channels, one of which sends the data through unchanged and the other of which goes to the pipeline for the AR algorithm, connect these two blocks. Before being output to a monitor, the real/virtual selector combines the two streams. Many fundamental picture processes (including edge detection and generalized convolution) were put into practice. Depending on the application, these operational units could be put together in a variety of ways to create the pipeline for the AR algorithm. Comparisons of performance between mobile devices and desktop computers revealed significantly higher efficiency in terms of clock cycles-per-operation. But because the FPGA runs at a slower clock speed (100 MHz as opposed to 2 GHz for the PC), some tasks took longer to finish in absolute terms [13].

The study of mobile augmented reality is one of the most significant developing research fields in the field, in contrast to the dearth of research into specialized embedded systems. The quick evolution of mobile phones and related gadgets is advantageous for researchers working in this field. Many phones now come equipped with a high-resolution camera and a 3D graphics accelerator, making them desirable platforms for vision-based augmented reality apps. The on-board graphics hardware is more likely to be present and is clocked at greater rates in current-generation applications processors. For instance, the previously stated OMAP3430, which has a C64x + DSP and a POWERVR SGX 2D/3D graphics accelerator, is capable of 600 MHz clocking.

AR apps now have additional capabilities thanks to embedded vision technologies. Real-world visuals are mixed or superimposed with virtual images in augmented reality (AR), whereas virtual reality is a totally immersive experience. Although embedded vision is essential to VR applications as well, AR is now a more economically feasible technology with potential applications across many different industries. Without embedded visual capabilities, augmented reality would not exist in its current form. There are standard best practices that every developer should adhere to when building an AR application. Primarily, computer-generated visuals need to be realistic and easy to understand, and they need to be aligned with the actual scene using virtual markers. Anything less than this will make AR ineffective. As real-world lighting and shadows change, reproduced light and shadows must likewise do the same. The image quality of virtual items must be on par with that of physical displays, and any positioning markers must be obscured to the user. While achieving all of this is no easy feat, cutting-edge embedded vision technology is assisting AR developers in developing useful, interesting, and realistic AR apps across numerous industries. The current iteration of AR technology is made possible by embedded vision. Embedded vision, a crucial step in the AR process, will help the technology advance as additional applications for it are developed.

4.3.3 Custom AR

In the history of AR research, embedded AR platforms have been extremely rare, with development primarily concentrating on PC-style systems. In addition to creating the virtual object overlay, the earliest known example of an embedded AR system in [14] utilized multiple TMS320C40 DSPs to implement object identification and feature point detection algorithms. The first usage of an FPGA for augmented reality was reported in [15]. This system could track things and create the augmented display, but it was made to function with a standard PC rather than on its own.

4.4 Augmented Reality Components

There are mainly three components, hardware, software, and remote servers.

4.4.1 Hardware Components

Displays, input devices, tracking, and computers are the main augmented reality hardware components. A smartphone, handheld device, pair of smart glasses, or head-mounted display could serve as the display. Either cameras or webcams are used as the input devices. The gyroscope and accelerometer were among the sensors. All of the specifications needed to operate an AR application are present on mobile devices.

4.4.1.1 Display Components

Three general categories can be used to classify augmented reality display devices: Head-up displays (HMD), portable displays, and spatial displays.

Head-Up Displays (HUD)

A head-up display (HUD) overlays the user's field of vision with both images of the actual and virtual surroundings when it is worn on the head or as part of a helmet. Mission-critical applications like flight controllers and weapon system dashboards were the primary uses for heads-up displays. On transparent panels installed in front of the pilot, important information is presented. Three major parts make up a typical head-up display: a computer, a projector unit, and a viewing glass (combiner) (symbol generator). By minimizing the focus shift for pilots, HUDs aid in improving situational awareness. Heads-up displays have been incorporated into new car designs more frequently.

Portable or Handheld Displays

Small computing devices having a display that the user can hold in their hands are used in handheld displays. Mobile phones, PDAs, and tablet computers are the three main categories of currently utilized handheld displays for augmented reality systems. Advanced computer vision techniques are now usable by anybody thanks to augmented reality frameworks like ARKit, ARCore, and MRKit. All you need for a variety of AR experiences in portable or mobile AR is a smartphone.

Spatial Displays

With the use of video projectors, optical components, holograms, radio frequency tags, and other tracking technologies, spatial augmented reality (SAR) allows users to view graphic data directly on actual things without having to wear or carry a display. These types of displays, made popular in recent years by the Star Wars, Minority Report, and Iron Man series, use light diffraction to create three-dimensional representations of objects in actual space. One of the biggest benefits of holographic displays is that no special equipment is needed to observe them.

4.4.1.2 Input Devices

For AR systems, there are many different kinds of input devices including wireless wristbands and gloves [16]. Google Sky Map on an Android phone needs the user to aim the phone in the direction of the stars or planets he/she wants to know the names of. In the case of smartphones, the phone itself can be used as a pointing device. The choice of input devices is highly dependent on the display option and/or the sort of application for which the system is being created.

4.4.1.3 Tracking

The two main types of AR tracking are vision-based tracking and sensor-based tracking. Optical, magnetic, acoustic, and inertial tracking are further categories for sensor-actuated tracking. The vision-based tracking are two types: Marker-based and markerless tracking [17]. Digital cameras, optical sensors, GPS, accelerometers, wireless sensors, etc. are examples of tracking equipment. Every one of these methods varies in accuracy depending on the kind of system being created.

4.4.1.4 Computers

Using desktop augmented reality (AR), 3D objects can appear to float in front of a screen on laptop or desktop PCs. There is some kind of computer in every augmented reality setup. The complexity of computing systems for augmented reality can vary from straightforward handheld devices like smartphones and tablets to sophisticated distributed systems and laptops, desktop computers, and workstation class workstations. In some circumstances, a portable computer is in contact with a potent server that may be situated far away [18].

4.4.2 Required Software

A computer program adds the virtual components, which can be enhanced by adding video, GPS information, or even music to make them seem even more lifelike. The AR-based applications incorporate specialized software like Unity 3D. Utilizing 3D software enables the overlay of a digital projection over a physical object. Computers can run this wide range of software and smartphone apps. The top augmented reality applications available right now are listed below.

- **SketchFab**

  A 3D modeling platform called SketchFab allows users to create, share, find, purchase, and sell 3D, VR, and AR content. It is also one of the most well-liked augmented reality programs because it is compatible with all operating systems.
- **Adobe Aero**

  Adobe is well regarded for producing some of the best art and design products, and Adobe Aero is no exception. Without knowing any code, realistic augmented reality experiences with objects like 4D cinema can be created with this software.
- **Masterpiece Studio Pro**

  One of the greatest tools for 3D modeling in VR is Masterpiece Studio Pro. We can directly draw, model, optimize, rig, skin, and animate in 3D and VR with a PC and headset for designs and concept art. Projects can also be exported for future development into animation and game engines. With only one click, it is possible to generate a full model using its highly sophisticated auto-rigger and auto-skin features. Masterpiece Studio Pro is a popular among artists and designers due to its simple user interface, mesh processing capabilities, texture painting, retopology, stamps, and kitbashing features.

- **Unity 3D**

  Users can construct games in a variety of contexts using Unity 3D tool. It offers an intuitive drag-and-drop feature and a major C# scripting API for the Unity editor and plugins. Any operating mode can be selected for the interface. Without learning any computer languages, it is possible to use the software. Developers can recycle the appearance of old information and remove seams from multiple materials from a single dashboard.

- **Vuforia Engine**

  The globe tracking, object recognition, and cloud recognition functions of the engine are the main components of Vuforia Engine. These capabilities enable developers to make engaging and immersive AR experiences. The engine can follow a user's movements in the real world and produce virtual content in accordance with those movements thanks to world tracking. This makes it possible for a more immersive and realistic AR experience. The engine can recognize and track particular items in the actual world thanks to object recognition.

- **AR Kit**

  With AR Kit, developers can use the tool's several potent capabilities, such as the ability to track the environment and the objects within it and produce realistic lighting and shadows. Using ARKit, it is also simple to include interactive components like menus and buttons in augmented reality experiences. Users may use the app to screen the model on any flat surface and interact with it using their iOS or Android devices thanks to ARKit and ARCore. We can also construct AR mobile apps using frameworks like ARKit and ARCore.

- **Oculus Medium**

  A digital sculpting program called Oculus Medium works with 6DoF motion controllers and virtual reality headsets. It is voxel-based. A voxel is a cube made up of a grid of grids, similar to how a pixel is a grid. Vertex data, which resembles 3D vectors in many ways, is used in traditional 3D rendering and 3D applications. Oculus Medium is more akin to 3D Photoshop compared to Adobe Illustrator/Flash and Adobe Photoshop.

- **ROAR Editor**

  A no-code platform called ROAR Editor is made for people without any programming knowledge. It takes just a few simple steps for anyone to build AR. All significant augmented reality capabilities, including video, audio, action buttons (CTA), photos, text, 3D models, and even supplementary product information, are available in the ROAR Editor.

4.4.3 Remote Servers

The data must be rendered and stored using cloud servers or a dedicated hosting platform. Virtual pictures and projections are downloaded from the web or cloud server and displayed via a smartphone in response to a request received from an AR source. Utilizing a hosted service or cloud is suggested because the amount of data collected during the process can be significant.

4.5 Relation of 5G/6G with AR/VR Systems

One of the innovative technologies poised to revolutionize how we engage with virtual worlds is 5G. Businesses all across the world will be able to access the VR environment more easily because to 5G's increased data streaming potential, decreased latency, and increased reliability. A high bandwidth Internet connection is necessary for the effective application of AR and VR technologies for the economy, commerce, and everyday life. Therefore, 5G's extremely low latency and high throughput are essential for allowing these technologies' applications at a satisfactory level. We must stream 12 K video in order to achieve VR that is equivalent to HD. The FTC's recommendations would suggest we need a minimum connection speed of roughly 225 Mbps—and possibly more to be safe—if we assume that it has nine times as much image data as 4 K and that there is no additional bandwidth overhead. To enable real-time user interaction in the immense environment, we need to take the help of 5G speed networking for AR/VR systems as AR/VR cannot be compressed (coding and decoding is time-consuming). Thus, the per user data rate needs to touch the gigabits per second (100 Mb/s as a 5G target). It is shown in Fig. 4.4.

Fig. 4.4 5G 6G hierarchy with AR/VR

With the use of 5G or 6G enabled AR/VR technologies in e-health care, we can eliminate time and space barriers through remote surgery and guaranteeing healthcare workflow optimizations. There may be some hindrance such as, high cost, lack of real-time tactile feedback, stringent quality of services (QoS) requirements, ultra-low latency, and mobility support. However, we still prefer to adapt this technology for e-health to inspire the concept of intelligent network. By enhancing 5G networks with higher reliability and higher bandwidth channels, 6G networks improve on them. Future 6G networks will feature massively connected nodes for ubiquitous AR/VR experiences, deep-sea and sky networks, and low-latency industrial IoT (IIoT) [19, 20].

4.6 Applications and Future Research Directions

4.6.1 Applications in AR

a. **AR in Medical Applications**

   Surgeons might employ augmented reality as a training and visualization aid. It will be able to collect 3D datasets of a patient in real time using non-invasive sensors like magnetic resonance imaging (MRI), computed tomography scans (CT), or ultrasound imaging. These datasets might then be combined and shown in real time with a visual of the actual patient. A doctor would essentially have "X-ray vision" within a patient as a result. When performing minimally invasive surgery, which lessens patient trauma by making few or no incisions, this could be very helpful. Surgery is more difficult because minimally invasive treatments limit the surgeon's ability to see within the patient [21]. Without the need for more extensive incisions, AR technologies could provide an interior view. In the operating room and may be useful for common medical visualization duties. MRI or CT scans cannot reveal certain traits that a surgeon can see with the unaided eye, and vice versa. Surgeons would have simultaneous access to both types of data thanks to AR. Additionally, this might serve as a roadmap for precise activities like showing where to make a hole in the skull for brain surgery or where to take a needle biopsy of a small tumor [22].

b. **AR in Classroom Education**

   In order for students of all age groups to thoroughly understand the lessons included in the subjects covered in their curriculum, AR is willing to collaborate with instructors and other educators, such as college professors. Since AR technology does not require a specific infrastructure, disciplines that are connected to physics, math, music, or psychology can be covered at a lower cost. For example, with the help of AR language assistant, Mondly, students may learn a new language like German, Spanish, or French anytime, anyplace, and without any additional equipment.

c. **AR in Entertainment Industry**

With augmented reality, entertainment gains a fresh perspective and the audience may take an active role in the presentation rather than just being passive consumers. AR has been successfully applied in PR and marketing campaigns for motion picture, television, and other media promotional campaigns for entertainment purposes. Typically, these have involved printed graphic or actual object recognition, in which computer or smartphone software recognizes a special sign. With the potential of AR/VR application development, it is possible to create immersive music performances and increase audience engagement at museums, athletic events, and amusement parks.

d. **AR in Navigation and Driving**

The user's location through GPS, the initial camera measurement, and the location of the object are only a few of the inputs used by AR-enabled navigation applications. They keep track of how things move. This data is gathered by sensors and connected to the IMUs of the moving objects. The cognitive load placed on drivers can be lessened by incorporating AR into the driving experience. They have specific information given to them rather than having to interpret the abstract information they see on the vehicle's display to what they see in the real world. The world is given additional context, and it becomes simpler to understand with a pointer indicating the precise turn they must make or the precise spot they must stop at.

$\Longrightarrow$ CASE STUDY OF SAILOR ASSISTANCE (Embedded System Architecture for Mobile Augmented Reality Sailor Assistance Case Study [23]).

This work produces a novel method for designing embedded and reconfigurable systems that are tailored to augmented reality, with three primary contributions. The first solution focuses on the association and adaption of various prior contributions to the AR domain in order to select and fully specify a set of algorithms for position and attitude estimation without the use of gyroscopes. The second is embedded system architecture, which introduces it as a quick and easy object processor (OP) tailored for mobile augmented reality. Since objects can be dispersed among a specified number of OPs, the design is versatile and especially well-suited for data reuse. Finally, the OP introduces a brand-new hardware-based pixel rendering technique (IPS) that fully complies with OpenGL ES light model standards. Several FPGA targets have been used to implement the entire design.

e. **AR in Travel "N" Tourism**

For users from other countries, an AR app serves as a 24/7 guide and helps remove linguistic obstacles. It improves consumer comfort and convenience for everything from arranging travel and lodging to discovering the top locations for shopping, dining, and entertainment. Smartify, ViewRanger, AR City, etc., are a few travel applications used by travel companies and accommodations to display augmented reality. Any of these apps will help to find what someone is genuinely looking for, whether trying harder to find transportation, petrol stations, ATMs, temples, or thrilled about perusing the brochures of agencies/travel firms offering accommodations at budget-friendly rates.

4.6.2 Future Research Directions

A common smartphone user could now experience AR thanks to the 2017 release of iOS11 and it is ARKit. Android users can access AR due to Google's ARCore software. Pokemon Go has demonstrated its enormous popularity, which is perhaps the first application of augmented reality that comes to mind. Future AR/VR technology will offer individualized, easily accessible, and beautifully designed experiences. There will soon be a platform change when these factors take effect. According to me, new AR glasses with LTE capabilities will be available in three years and will displace smartphones. Given the market's recent volatility, some businesses might be holding off on taking any action. Companies may want to think about collaborating with knowledgeable vendors who can provide end-to-end product development with comprehensive engineering capabilities in order to successfully overcome the challenges associated with building out augmented and virtual reality technologies while maintaining market expectations and time-to-market. Nearly 90% of businesses already count on vendors to support their AR/VR needs, from creating planning to provide full production capabilities. Employing outside expertise and virtual and augmented reality technologies allows businesses to concentrate on their core competencies while providing a completely immersive experience [24]. Finding specific future study areas is a natural result of analyzing current and prior projects based on educational level, domain, tracking, cooperation capacity, agents, and interaction level. The suggested requirements for AR application design include flexible material that the teacher can modify in accordance with the needs of the students, guidance during exploration to optimize learning opportunities in a constrained amount of time, and attention to curricular needs [25].

- The study demonstrates that AR has been evaluated and found to be equally successful at all three educational levels: primary and secondary education, high school, and university education (university). Additionally, there is a tendency toward the use of AR in medical education, although little attention is paid to the technical or vocational evaluation of its usage in the classroom.
- The majority of research use augmented reality (AR) for alphabet learning, vocabulary learning, or early level science topics at the primary and elementary levels of education. It has been utilized as a learning resource to supplement STEM disciplines at the secondary (high school) and tertiary (university) levels. Future prospects in STEM are growing thanks to immersive learning technology. The usage of AR technology can be a useful resource for students to teach those subjects or skill training where actual content is not accessible or possible in a classroom setting.
- The majority of large-scale research have experimented with control and experiment groups. In comparison to traditional learning tools, the studies' main goal is

to raise students' attention, make them more relevant to the study themes, and give them greater confidence and happiness. Like EARLS, certain studies are assessed on a broad scale.

- Applications for location-based learning have made use of GPS. Most research uses Vuforia as their tracking SDK. The other primary tracking libraries include ARKit, ARCore, AR Foundation, MRTK, and AR.js. These work with a brand-new lineup of expensive gadgets. Numerous unique solutions are also employed.
- Desktop PCs have traditionally been the primary platform for AR applications, followed by smartphones and head-mounted displays (HMDs). Due of its portability, tablets and smartphones are currently gaining more users as a result of the saturation. Particularly, the availability of personal gadgets has increased the likelihood of utilizing a smartphone. Due to their affordability—good HMDs are significantly more expensive than smartphones—tablets and smartphones are the most dependable devices for augmented reality in education. The traditional desktop is shifting further in the direction of smartphones and HMDs like Magic Leap and Hololens.

However, it is expected within the year 2045 that VR technology graphics will be photo-realistic, display will be 360° immersion, interaction should be full gesture based, navigation will be natural and use by millions of users. Future research can be done in each of these below areas.

- Combines real and virtual images as display.
- Registered in 3D tracking technologies.
- Real-time interaction.

Since augmented reality is still in its early stages, the potential uses are endless. Advanced research in augmented reality (AR) involves creating controlled environments with a variety of sensors and actuators as well as using head-mounted displays and virtual retinal displays for visualization [26]. From real estate to tourism, virtual reality is demonstrating some useful business use cases. These cutting-edge VR experiences are emerging in a variety of global sectors. Since a few years ago, smartphones and headsets have been the most popular method of delivering a VR experience. The easiest way for consumers to start using VR has been through this. Compared to the market for augmented reality today, virtual reality is far more developed. To create an immersive VR experience, the necessary hardware platforms and software tools are now available. Virtual reality experiences are rapidly expanding their applications in our lives thanks to the availability of increasingly sophisticated systems like the Oculus Rift and 360-degree cameras. Due to its limits, lack of standards, and higher price, augmented reality technology is still in its infancy compared to virtual reality, but it is already being used in sectors including manufacturing, healthcare, and logistics. Experiences with augmented reality are often provided through headsets like Meta, ODG, Vuzix, and HoloLens, and early indications indicate that the technology is about to revolutionize business and industrial markets. Before these technologies are widely used, however, there is more work to be done.

The AR market is expanding rapidly. Experts predict that the sector would be valued more than $76 billion by 2030, up significantly from $26.75 billion in 2021. The inventions that emerge over that time will be intriguing to see, as will how much simpler augmented reality can improve our lives. One such cutting-edge technology that makes science fiction a reality is unquestionably augmented reality. Holograms are now all around us in the real world, just like in the Star Wars and Marvel films, creating a unique immersive experience that goes beyond simple entertainment. Today, augmented reality is a powerful tool for business. Augmented reality is employed in a variety of different industries, including retail, commerce, gaming, healthcare, and even the military, to address a variety of business difficulties. To understand the direction the industry is go, it is critical to keep an eye on these technologies. Consider how these below augmented reality ideas that will be prevalent in 2023 can help your own company as we go over them.

(a) **Enter the Metaverse**

You probably do not find it surprising that augmented reality is being used in conjunction with other metaverse technologies. Since Facebook changed its name to "Meta", the metaverse has bombarded the news media. It is not just marketing nonsense, though. Dismantling the partitions between the virtual and real worlds is one of the objectives of metaverse technologies. Due to the ability of augmented reality to show virtual items that are integrated into our real world, both consumers and companies can benefit from this technology. AR is a terrific place to start if we want to integrate digital experiences into the physical world. Businesses are already developing camera filters that do this using body and facial tracking, as well as sophisticated scene depth sensing. Ready Player Me and Geenee AR together to make this experience a reality. You may practically "wear" your avatar on camera by adding it to Geenee's WebAR Builder program. Your Ready Player Me character's cosmetic accessories, including NFT accessories, are also taken into account by the software. This innovation is nothing new. It has long been observed being used with apps like Snapchat and Instagram. However, the app's revolutionary feature is the way users can import their existing avatar from other platforms and use it in augmented reality. This technology may be applied in the future to better hybridize online meetings. An AR avatar of the individual could represent them in your meeting if one member of your team is using a VR headset and the rest of you are not.

Spatial audio is crucial for boosting the immersion of AR experiences, despite the fact that it may not initially appear to be an augmented reality technology. The inclusion of all of our five senses in the process is something that metaverse technologists are preoccupied with, and this includes hearing. 3D audio is required to increase the immersion of VR and AR experiences. Based on their own position, users should be able to determine where a sound is coming from in 3D space. Recently, Meta upgraded its AR Spark Studio with a sophisticated engine that combines various sounds to produce sound effects. This enables developers to produce multisensory effects that let users immerse themselves in the augmented reality experience using both sight and sound. We may respond to user interaction with our AR effect by making sounds play.

(b) **AI Inspired Augmented Reality**

Artificial intelligence and augmented reality work two ways well together:

- Software for facial and spatial recognition, which is essential for AR, is powered by artificial intelligence.
- Innovative solutions can be provided by combining AR and AI technologies.

These positions do not have to be mutually exclusive. They frequently meld into one another. Artificial intelligence and augmented reality are two distinct fields of technology. However, given AR's requirements, it is not surprising that AI and AR get along well. Environment sensor data must be interpreted using complex algorithms. In comparison to a model created solely by humans, AI can streamline that process and increase accuracy. The application ClipDrop is an illustration of this in action. With the help of the app, users can instantly turn a physical thing into a 3D object for usage in applications like PowerPoint, Photoshop, Google Docs, and more. Real-world objects can also be imported into metaverse settings using 3D scanning. Businesses might find that 3D scanning is a terrific method to expedite the process of offering products for virtual trial experiences. A further application of merging AI with AR is automatic design. An example of this technology in use is the app SketchAR. Using this software, users can doodle at will in AR. They can also employ an AI to draw for them, though. The AI can create structures swiftly. This demonstrates that using the real world as the source environment, AI systems are capable of designing objects in 3D space. This may indicate that AI will eventually be able to design and produce buildings for use in the actual world.

(c) **Mobile Communicated Augmented Reality**

Mobile devices have been one of the key platforms for distributing augmented reality experiences. The majority of consumers own a mobile device, and consumer use of AR headsets is still in its infancy. As a result, companies have discovered several ways to employ mobile devices for augmented reality. Google unveiled a new geospatial experiences API in 2022. This makes it possible for developers to design experiences linked to particular astronomical places. AR encounters in the past have only been relative to the user or in arbitrary places chosen by the user. WebAR is a significant augmented reality movement. Powered by web browsers, WebAR does not require users to download additional software. The ideal situation for accessibility is this. The most basic AR experiences are available through WebAR, which lacks many of the features that native AR on mobile devices can provide. WebAR, however, has its uses, especially for straightforward experiences and includes changing the color of objects or hair, replacing the background, and creating simple 3D creations. WebAR enables easier virtual try-on experiences. Many companies, like L'Oréal and Maybelline, use them for their cosmetic goods.

(a) AR navigation is easier to use and more feasible than ever in 2022. Most notably, indoor navigation is now much more practical than in previous years thanks to the development of technologies like Bluetooth Low Energy (BLE) antennae,

WiFi RTT, and ultra-wideband (UWB). Displaying AR directions in big interior spaces like distribution centers, shopping malls, and airports is one of this technology's most practical uses.

(b) By providing hands-free solutions offered by MR/AR and IA-based diagnostic tools, augmented reality and AI will, according to Deloitte Research, alter the conventional healthcare business paradigm. For instance, the Microsoft Hololens 2 can give the surgeon information while letting them to utilize both hands for the procedure. The adoption of augmented reality technology is essential to solve challenges like the difficulty of remote patient support and the growing strain on hospitals as a result of the continuous restrictions related with COVID-19. This covers both mental health apps and telesurgery options that assist people in maintaining psychological balance during these trying times. For instance, tools like drawing and annotation capabilities on the 3D screen can greatly improve and simplify doctor–patient communication. When combined with machine learning algorithms, augmented reality technology can be a useful tool for identifying diseases. For the Department of Defense (DoD), Google developed an AR-based microscope back in 2020 to increase the precision of cancer diagnosis and therapy. A hybrid gadget like this one employs a camera to take real-time photos that are processed using computer diagnostics to deliver results right away and detect diseases in their early stages.

(c) The automotive business may benefit from a variety of distinct augmented reality applications. One of the more intriguing and upcoming technologies in this field is augmented reality, which uses a heads-up display to highlight on-road objects (HUD). This can help keep drivers' eyes on the road while also alerting them to risks and GPS directions. In addition to providing entertainment and information, augmented reality is used in applications like 3D automobile manuals.

4.7 Conclusion

The possibilities with augmented reality are infinite, as the technology becomes more widely used and more reasonably priced. Anyone can anticipate seeing more immersive and interactive ways of experiencing and consuming material when combined with other extended reality technology. In this chapter, our discussion is mainly focused on the various tools, technologies of AR/VR systems architecture including the applications on various domains. This chapter also gives some lights on the future research directions of this AR/VR embedded system where researchers can do an extensive research.

References

1. Lee J, Kim Y, Heo MH, Kim D, Shin BS (2015) Real-time projection-based augmented reality system for dynamic objects in the performing arts. Symmetry 7(1):182–192
2. Toledo J, Martınez JJ, Garrigós J, Toledo-Moreo R, Ferrández JM (2010) Design of embedded augmented reality systems. Augmented reality
3. Milgram P, Kishino F (1994) A taxonomy of mixed reality visual displays. IEICE Trans Inf Syst E77-D:12
4. Ismail AW, Sunar MS (2009) Collaborative augmented reality: multi-user interaction in urban simulation. Visual informatics: bridging research and practice, pp 382–391. https://doi.org/10.1007/978-3-642-05036-7_36
5. Cipresso P, Giglioli IAC, Raya MA, Riva G (2018) The past, present, and future of virtual and augmented reality research: a network and cluster analysis of the literature. Front Psychol 2086
6. Diguet JP, Bergmann N, Jean-Christophe M (2013) Embedded system architecture for mobile augmented reality. Sailor assistance case study. In: 3rd international conference on pervasive and embedded computing and communication systems (PECCS)
7. Kisacanin B, Bhattacharyya SS, Chai S (eds) (2008) Embedded computer vision. Springer
8. Saha S, Bhattacharyya SS (2009) Design methodology for embedded computer vision systems. In: Embedded computer vision, pp 27–47. Springer, London
9. Neuvo Y (2004) Cellular phones as embedded systems. In: 2004 IEEE international solid-state circuits conference (IEEE Cat. No. 04CH37519). IEEE, pp 32–37
10. Bouju A, Stockus A, Bertrand F, Boursier P (1999) Client-server architecture for accessing multimedia and geographic databases within embedded systems. In: Proceedings. Tenth international workshop on database and expert systems applications. DEXA 99. IEEE, pp 760–764
11. Lampropoulos G, Keramopoulos E, Diamantaras K (2022) Semantically enriched augmented reality applications: a proposed system architecture and a case study. Int J Recent Contrib Eng Sci IT 10(1)
12. Longobardi A, Tecchia F, Carrozzino M, Bergamasco M (2021) Efficient augmented reality on low-power embedded systems. In: International conference on augmented reality, virtual reality and computer graphics. Springer, Cham, pp 227–244
13. Shinde GR, Dhotre PS, Mahalle PN, Dey N (2021) Augmented reality and IoT. In: Internet of things integrated augmented reality. Springer, Singapore, pp 55–71
14. Uenohara M, Kanade T (1995) Vision-based object registration for real-time image overlay. Comput Biol Med 25(2):249–260
15. Luk W, Lee TK, Rice JR, Shirazi N, Cheung PY (1999) Reconfigurable computing for augmented reality. In: Seventh annual IEEE symposium on field-programmable custom computing machines (Cat. No. PR00375). IEEE, pp 136–145
16. Reitmayr G, Schmalstieg D (2003) Location based applications for mobile augmented reality. In: AUIC2003
17. Ashwini KB, Patil PN (2020) Tracking methods in augmented reality–explore the usage of marker-based tracking. In: Proceedings of the 2nd international conference on IoT, social, mobile, analytics & cloud in computational vision & bio-engineering (ISMAC-CVB 2020)
18. Craig AB (2013) Understanding augmented reality: concepts and applications
19. Sharma SK, Woungang I, Anpalagan A, Chatzinotas S (2020) Toward tactile internet in beyond 5G era: recent advances, current issues, and future directions. IEEE Access 8:56948–56991
20. Mukherjee A, Panja AK, Dey N, Crespo RG (2022) An intelligent edge enabled 6G-flying ad-hoc network ecosystem for precision agriculture. Expert Syst e13090
21. Azuma RT (1997) A survey of augmented reality. Presence: Teleoper Virtual Environ 6(4):355–385
22. Shinde GR, Dhotre PS, Mahalle PN, Dey N (2021) Internet of things integrated augmented reality. Springer
23. Diguet JP, Bergmann N, Morgère JC (2015) Dedicated object processor for mobile augmented reality-sailor assistance case study. EURASIP J Embed Syst 2015(1):1–17

24. Iqbal MZ, Mangina E, Campbell AG (2022) Current challenges and future research directions in augmented reality for education. Multimodal Technol Interact 6(9):75
25. Liu S, Manocha D (2022) Future research directions. In: Sound synthesis, propagation, and rendering. Springer, Cham, pp 65–67
26. Carmigniani J, Furht B, Anisetti M, Ceravolo P, Damiani E, Ivkovic M (2011) Augmented reality technologies, systems and applications. Multimed Tools Appl 51(1):341–377

Part II
Biological Applications of Data Analytics

Part II
Biological Applications of Data Analytics

Chapter 5
Computational Biology Toward Data Analysis

5.1 Introduction

We are currently living in an era of technology where computers and computing have always had an impact on our lives. People do their internet transactions through online banking, online shopping, and online stock trading. The computation on supercomputers can be used to learn about weather forecasts and natural disaster predictions. We also discover a significant number of computers and computing applications used in biological and medical research, which led to the development of the field, of computational biology. It is a process that helps to develop new hypotheses, identifies new patterns, and finds novel functional elements using a comprehensive, data-driven, genome-wide, and systematic approach. It aids in the correction of flaws in standard experimental biology methods. The best strategy to learn biological data is supposed to use a strong combination of computational and experimental methods. In terms of creating biological data, the last two decades have seen extraordinary technological advancement: Microarrays, sequences, protein–protein interaction, and other high-throughput methods have resulted in an explosion of data.

5.2 History of Computational Biology

Computational biology has a long and illustrious history. The use of computational approaches to protein sequence analysis laid the groundwork for computational biology in the early 1960s [1]. Physical chemist Margaret Dayhoff was a pioneer in the use of computational techniques in the study of biochemistry. Dayhoff has made such a significant contribution to this discipline that David J. Lipman, a former NCBI director, has referred to her as “the mother and father of bioinformatics”. In 1962, Dayhoff and Ledley introduced the COMPROTEIN algorithm, which divides proteins into smaller parts and then assembles the pieces to make a whole protein sequence that runs on IBM7090. Additionally, she released the “Atlas of Protein Sequence and

S. Chakraborty and L. Dey, *Computing for Data Analysis: Theory and Practices*,
Data-Intensive Research, https://doi.org/10.1007/978-981-19-8004-6_5

Structure", the first computational biology database. Later, the database was taken over by UniProt following Margaret Dayhoff's untimely death, and it became the Protein Information Resource (PIR). Initially, the database contains only 65 protein sequences in the year 1965. Today, it contains around 43 million protein sequences. Therefore, it became challenging to analyze this huge number of proteins and continue utilizing the manual alignment method for calculating sequence similarity. The Protein Data Bank (PDB), which was established in 1977, was designed to store the three-dimensional protein and nucleic acid structures. Researchers are beginning to anticipate protein structures using computational methods at the same time. As a result of the critical mass of scientists, researchers, and technological advancements in the year 1980, the interdisciplinary area of computational biology quickly expanded.

The Genbank database was established in 1982 to preserve nucleotide sequences, and it has since grown rapidly. Nearly 31,000 nucleotide sequences have been solved as of the end of 1989. Since there were so many protein sequences in this database, the available methods at the time were unable to process it quickly enough. The human genome project started in 1986 and was completed in April 2003. It generated the entire genome sequences in humans involving 3 billion bases of DNA. The database is available at the TIGR database (http://www.tigr.org/tbd/). This information paved the way for the diagnosis of diseases and creates potential drug targets for new diseases. Although the human genome project was completed in April 2003, the exact genome is still unknown. Microarray technology and expressed sequence tag sequencing were developed and became widely employed in the 1990s. The Sequence Read Archive was created to hold the read sequences after next-generation sequencing technologies were introduced in 2005. At this time, there are more than 1800 trillion nucleotides stored in the database.

Until recently, gene prediction, sequence alignment, and protein–protein interaction prediction were determined by carrying out laboratory-based experiments that can identify only a small number of specifically targeted interactions. However, these methods have shortcomings. Not only labor-intensive, time-consuming, and costly, they produce a lot of false positive and false negative outcomes. As the human body is composed of trillions of cells, the analysis of genes and interactions among them will take more time compared to other organisms. Therefore, computational methods become a better alternative that can predict a large set of outputs within a small time.

Various approaches for predicting DNA and RNA sequencing, protein–protein interaction (PPI) prediction, and microarray data analysis have evolved over the years. The available computational methods can be divided into four main categories: network-based approach, genomic and structural information-based approach, machine learning-based approach, and text mining and literature mining-based approach. In this chapter, we will discuss the numerous genomics and genetics issues that have been tackled with the use of machine learning techniques.

5.3 Biological Data Types

Computational biology deals with the application of mathematical tools to extract useful information from large amounts of complex and high-dimensional biological data. Biological data refers to the information gathered from the biological organism. The human body is a complicated system made up of cells with chromosome-containing nuclei. The DNA-based genes on chromosomes contain protein-coding information. Typically, DNA creates RNA, and RNA creates protein. This is a complex procedure, and it is called the central dogma of biology.

It consists of two steps, as shown in Fig. 5.1. Transcription, the first stage, involves turning DNA into messenger RNA (mRNA), while translation, the second step, involves turning mRNA into protein. Now, gene expression is the procedure through which a gene is translated into a product, such as a protein or RNA. The composition and operation of a cell are greatly influenced by the level and structure of its proteins. It is important to analyze the level of proteins and gene expression in cells to understand the cell's biological processes. Therefore, computational biology mainly focuses on the analysis of the five types of data, like DNA/RNA sequence, protein structure, protein sequence, gene expression, and pathways (shown in Fig. 5.2) to investigate medical genetics and the diagnosis of diseases.

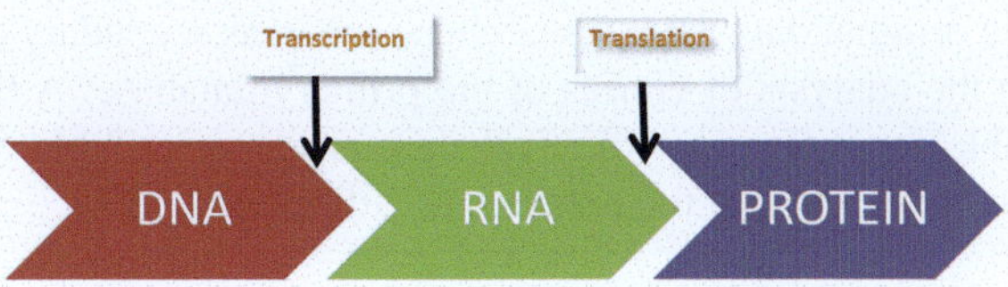

Fig. 5.1 Diagrammatic representation of central dogma

Fig. 5.2 Biological data types

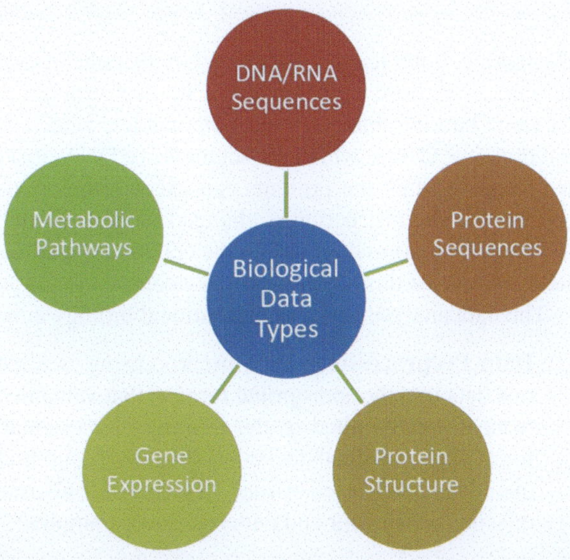

5.4 Biological Databases

Biological data is interrelated. For example, while dealing with genome sequences, we have to keep in mind the structure, function, and amino acid sequences. Apart from that, biological data formats are extremely complicated, and traditional database management systems (DBMS) do not adequately address the need for complex data structure as compared to most other applications. This leads to the concept of biological database to store the biological data in an electronic format. It is a collection of searchable, structured, and periodically updated data. Biological databases can essentially be divided into the following groups based on the sorts of data stored in them: (1) DNA, (2) RNA, (3) protein, (4) expression, (5) pathway, (6) gene ontology. There are different biological databases that contain different biological data. For instance, DNA information can be found in nucleic acid databases, gene-level information can be found in genomic databases, protein information can be found in protein databases, and categorization information about proteins can be found in protein families, domains, and functional sites. These databases serve as repositories of biological data for researchers. Each entry in the database contains information about the nucleotide sequence, protein sequence, 3D structure, etc.

Biological databases can be categorized into two groups: primary databases and secondary databases, depending on the source and caliber of the content. Primary databases include original data from researchers, whereas secondary databases contain results from the analysis of entries of the primary database. A list of database resources available for different biological data is given in Table 5.1. A defined algorithm is required to analyze the contents of a database.

5.5 Data Analysis

The basic cellular mechanisms of all living cells of every living organism are mostly the same. All those cells may differ in their structure and complexity but consist of the same building blocks—deoxyribonucleic acid (DNA), ribonucleic acid (RNA), and proteins. Genes, which are segments of DNA, inside a cell, raise the instructions to proteins so that the proteins can execute the appropriate functions for the cell. However, if the instruction is changed or mutated, the protein function is changed and leads to abnormal growth in the cell. Computational biology is used to speed up the discovery of this abnormality in a cell.

The data analysis using computational biology consists of the following steps:

1. **Data Preprocessing**: Data preprocessing is a technique for transforming the raw data into the appropriate form so that valuable information may be extracted from it and supplied to the training of particular machine learning problems for effective medical decisions, diagnoses, and treatments. Raw datasets are frequently characterized by inaccuracies while also being inconsistent, out of range, missing in behavior and patterns, and incomplete. To handle missing values and

Table 5.1 List of biological database types, names, their descriptions, and web interfaces

Database type	Database name	Descriptions	Web interfaces
Primary database	GenBank	Nucleotide sequence database	http://www.ncbi.nih.gov/GenBank
	DDBJ	Nucleotide sequence database	http://www.ddbj.nig.ac.jp
	PIR	Protein Information Resource	https://proteininformationresource.org/
	GEO	Functional genomics database	http://www.ncbi.nlm.nih.gov/geo/
	PDB	3D structure of protein database	http://www.rcsb.org/pdb
	EMBL	Molecular biology database	http://www.ebi.ac.uk/embl.html
Secondary database	CATH	Hierarchical classification of protein domains	http://www.cathdb.info
	OMIM	Online Mendelian Inheritance in Man	http://www.ncbi.nlm.nih.gov/
	Pfam	Protein family database	http://www.sanger.ac.uk/
	KEGG	Regulatory pathways of genomes	http://www.genome.jp/kegg/pathway.html
	SCOP	Structural classification of proteins	http://www.scop.mrc-lmb.cam.ac.uk/
	Prosite	Database of protein domains, families, and functional sites	http://www.expasy.org/prosite

deal with discrepancies, preprocessing is necessary. Data discretization, parsing, reformatting, conversion, and storage optimization are performed in this stage.

2. **Feature Extraction**: The procedure of selecting a subset of features that will most effectively describe the data required for analysis is known as feature extraction. Finding the most condensed and informative set of features (distinct patterns) is the goal of feature extraction in order to improve the effectiveness of the classifier of ML algorithms.
3. **Modeling**: This step helps to automatically handle and analyze biological data that are created using data preparation and filtering. It includes different classification techniques, statistical methods, clustering, and reverse engineering on biological networks.
4. **Evaluation**: Model evaluation is the practice of employing several evaluation measures to comprehend the performance and strengths and weaknesses of a computational model. In the early stages of research, it is crucial to evaluate a model's effectiveness. The model evaluation also aids in model monitoring. Three factors—probability, possibility, and plausibility—are used to assess a model. A workflow of data analysis in computational biology is shown in Fig. 5.3.

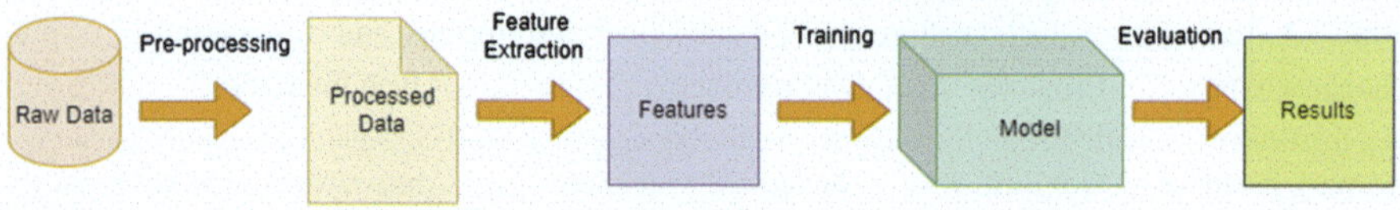

Fig. 5.3 Computational biology workflow in data analysis

5.5.1 DNA/RNA Sequence Data Analysis

A biological macromolecule is DNA, also known as deoxyribonucleic acid. It is made up of two strands that have been coiled around one another. It carries the genetic material for life and directs how life functions [2]. DNA may be thought of as the design manual for cellular equipment. During an organism's existence, RNA aids in carrying out this plan. Adenine (A), guanine (G), cytosine (C), and thymine (T) are the four nucleotides that make up DNA. The typical example of a DNA sequence is .... ATCGCTGATGA...., which is a lengthy but limited string made up of the letters A, C, T, and G in the "nucleotide" alphabet. Therefore, DNA sequencing is the process of determining these nucleotide sequences of a given DNA molecule [3]. RNA is single-stranded, in contrast to DNA. It has ribose instead of deoxyribose and uracil (U) in place of thymine (T). DNA/RNA sequencing has become a new advanced technology for medical research by contrasting normal and mutant sequences. Researchers can diagnose different diseases like cancers using sequences, and it can be used to guide patient treatment.

Sequencing technology has gone through many stages of development along with biological information technology. The first-generation sequencing technology was the chain termination approach proposed by Sanger. For conventional sequencing applications and verification, Sanger sequencing is still widely utilized today. However due to its high sequencing cost and limited throughput, it is not currently used on a really massive scale. Accuracy and speed of identification are equally as crucial as understanding DNA sequence. There are millions of genes in a DNA sample. As a result, classifying this is challenging. In order to improve the data processing capabilities of DNA and RNA sequencing and produce useful biological information, machine learning has been extensively applied in sequence data analysis [2, 4, 5].

In [6], Gunasekaran et al. used K-mer encoding with different machine learning (ML) algorithms to classify DNA sequences of COVID-19 infected and healthy genome samples. They have achieved 93.13% accuracy on the test dataset. Using linear classifiers, the complete and partial genomic sequence of the HCV dataset is predicted [7]. Pablo Millán Arias et al. presented a novel deep learning method for the unsupervised clustering of DNA sequences [8]. With accuracies ranging from 77 to 100%, the DeLUCS clusters learned from huge and diverse datasets match real taxonomic groups. Rashed et al. described a customized ML model for global sequence alignment in [5] and achieved an accuracy of 99.7% using a multilayer perceptron on two real DNA sequences with a length of 4.1 M nucleotides. Two commonly used DNA sequence alignment algorithms based on a lookup table (LUT) were utilized, and a modified convolution neural network model achieved 98.3% accuracy in [9]. In this chapter, we have considered three uses of ML in DNA sequence data analysis: aligning DNA sequences, classifying DNA sequences, and clustering DNA sequences.

5.5.1.1 DNA Sequence Alignment

Two biological sequence characteristics can be arranged to find regions of similarity or compare the dissimilarity using a technique called sequence alignment. It is among the most crucial elements of bioinformatics. It is necessary for phylogenetic analysis, comparative genomics, and the prediction of protein structure and function in order to hunt for conserved areas. To accommodate the diverse needs of biologists, a multitude of techniques and tools for sequence alignment have been developed.

Calculating the number of matching symbols, for instance, is the easiest technique to compare two sequences of equal length. The alignment score of two sequences is a number that expresses how similar two sequences are to one another. Figure 5.4 shows an example of two sequences of equal length with four non-matching characters. There are two types of sequence alignment: double sequence alignment and multisequence alignment. Double sequence alignment was the starting point for early research on biological sequence alignment. Needleman and Wunsch created dynamic programming approaches for double sequence alignment based on how similar the entire sequence was. Then, the double sequence alignment is expanded upon by multisequence alignment. With FASTA and BLAST, expected sensitivity is reduced in exchange for faster processing. The most popular and efficient multiple sequence alignment algorithm is CLUSTALW. The key issue with sequence alignment is determining if the algorithm's sensitivity to alignment and its effectiveness for sequences with significant variations have been improved. Sequence alignment for substantially big input sequences is a challenging process. Therefore, machine learning is widely used for analyzing large-scale sequences.

In [10], Zhao et al. used an improved ant colony algorithm for efficient and feasible DNA sequence alignment. The method resolves the issue that the outcome falls into a locally optimum solution by tweaking the beginning and end positions of the ants and altering the pheromone at various intervals, however, the quantity of calculation is huge, and the process is time-consuming. DNA multisequence alignment method MWPAlign, based on graphs, was proposed in [11]. This approach transforms the sequence alignment issue into the graph's greatest weight path by expressing sequence information as a structure graph. Stochastic approaches are among the heuristic strategies that have received the most attention in research to address multisequence alignment issues. Therefore, a genetic algorithm is used in many studies for DNA multiple sequence alignment.

ATCGTTCATTACGTA
ATCGATCATCGCGTT

Fig. 5.4 Example of two sequences with four non-matching characters, highlighted with red color

5.5.1.2 DNA Sequence Classification

Sequence classification helps identify genes in DNA molecules by predicting the kind of DNA sequence based on how similar its structure or function is to that of other sequences. The DNA sequence is difficult to comprehend and analyze since it is incredibly lengthy, with a maximum length of 32,000 nucleotides. Therefore, computational techniques are vastly used to classify the partial and complete genomic sequence in different literature.

Convolutional neural networks (CNNs), a deep learning model type that can extract characteristics of high-level abstraction from data with the least amount of preparation, have gained popularity in recent years for DNA classification. CNN has also been used to solve text data issues including topic classification, spam detection, and sentiment analysis. As DNA sequences are also text data, many researchers convert DNA sequences to word sequences with position information to apply the CNN model [12]. A one-hot vector of size 64 then serves as the representation for each word. Then, it is possible to create a two-dimensional numerical matrix that contains details on the precise placement of each nucleotide in the sequence. As classification is a supervised learning method, the matrix also contains the labels of the training data in order to predict the category of unknown new samples. The CNN model that performs the classification is then employed with this matrix as input. Finally, the performance of the classifier is evaluated with the help of accuracy, precision, and recall factors.

5.5.1.3 DNA Sequence Clustering

One of the machine learning techniques that are most frequently employed is cluster analysis. In contrast to classification, in clustering, we are not aware of the precise categories in advance. Sequence similarity analysis is the foundation for DNA sequence clustering. DNA sequences with comparable traits are grouped together in a cluster using cluster analysis, and the roles of the biological sequences are then examined. Researchers may use the data gathered through sequence clustering to identify gene alterations, connections with phenotypes and illnesses, and possible therapeutic targets. Many studies in DNA sequence clustering are now being conducted that are based on the regional DNA properties. Graph-based algorithms, K-means, and hierarchical clustering are widely used in DNA clustering. Apart from machine learning applications, a lot of tools for sequence clustering, namely CD-HIT, UCLUST, and DNACLUST, based on greedy algorithms are available [13]. The issue with greedy techniques is that they are sensitive to the choice of initial sequences.

The algorithm of DNA sequence clustering can be discussed as follows:

1. At first, the DNA sequence is broken into k-mers. A k-mer is gathered from every feasible point in these sequences of k continuous symbols; for instance, the sequence ATGCGTAA would be broken apart, into the ATGCG, TGCGT, GCGAT, and CGTAA 5-mers. The size of the input sequences affects the choice of this *k* value.

2. The target sequence is split into k-mers, which are then combined, converted into random vectors of a predetermined length, stripped of any positional information (such as positions 0, 1, 2, and 3), and added. This method is known as random indexing [14].
3. The clustering is used on the signatures in the next step once the sequences have been transformed into k-mer histograms.
4. For K-means clustering, choices can be made by initializing k unique sequences at random and designating them as the cluster centers. The next two processes which can be repeated several times to arrive at locally optimal solutions are:
   (a) Each sequence is placed into the cluster with the closest cluster center after iterating over all of the sequences.
   (b) After going over each cluster, recalculate the cluster center to make it equally distant from each member of the cluster.
5. Each sequence is regarded as belonging to the cluster it was placed at last after the specified number of iterations has been completed.
6. Finally, the clustering results are evaluated using the parameters like inter-cluster similarity, intra-cluster similarity, silhouette score, purity, normalized mutual information (NMI), the time required, and memory utilization. Different tools for clustering biological sequences are also used for comparison purposes.

5.5.2 *Microarray Data Analysis and Preprocessing*

The term "gene chips" is frequently used to refer to DNA microarrays, also known as DNA arrays. A unique piece of glass is used to construct DNA microarrays, which have a microarray. It comprises tens of thousands to hundreds of thousands of nucleic acid probes in a few square centimeters. The probes are coupled with the DNA, RNA, and complementary DNA (cDNA) in the specimen, and they are then detected by a fluorescence or current. The fundamental idea behind a microarray is the hybridization of two DNA strands, a phenomenon that links together nucleic acids by forming hydrogen bonds between matched nucleotide base pairs. To understand the genome, beyond its static view, we have to use the DNA microarrays, which enable us to measure thousands of genes simultaneously. A thin membrane carrying samples of thousands of genes arranged in some regular manner makes up this instrument for analyzing gene expression. This technology is used to identify the sequence of genes and to determine the expression level of genes. It combines statistical analysis with wet-lab methods. A diagrammatic representation of microarray gene expression data is depicted in Fig. 5.5. A gene is represented by each row. An experimental condition is represented by each of the columns (a sample, a time point, etc.). Many different industries, including biotechnology, agriculture, food, cosmetics, and computers, employ microarrays. The expression levels of thousands of genes, the connections between them, their activities, and the classification of genes or samples may all be monitored and studied simultaneously using this technique. The expression levels

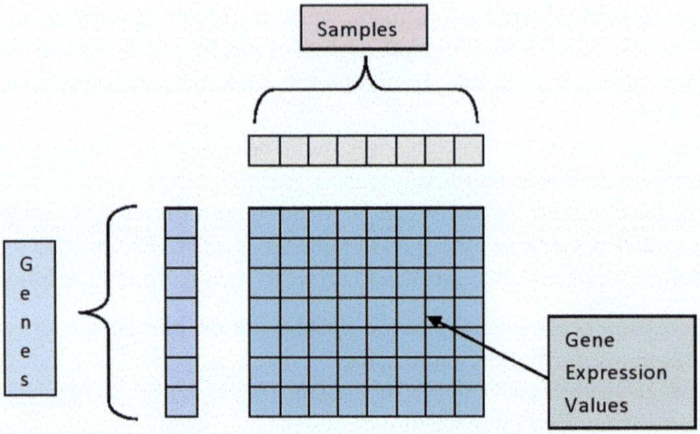

Fig. 5.5 Diagrammatic representation of the microarray gene expression data

can vary according to the experimental conditions, the environment, medicine, a condition, etc. Therefore, gene expression profiling can aid in the discovery of drugs, the impact of changing environmental factors, and the distinction between a diseased and healthy states.

Numerous research have employed microarray data to comprehend the pathophysiology of various illnesses [15]. Transcription and translation are the first two steps in the cell regulatory process for normal tissue. It offers the expression levels of thousands of genes at a particular time and circumstance in both the tumor and normal cells. When there is a statistically significant difference in expression levels between two experimental situations, such as a disease state and a healthy state, this is known as a differential gene expression (DGE) [16, 17]. Networked Gene Prioritizer (NGP), a technique created by Wu et al. in [18], was used on many datasets related to breast cancer and lung cancer. In this analysis, the scientists discovered several possible key genes. In [19], a biometric analysis is carried out on data from several cancer microarrays. They have demonstrated that PPIs from all cancer types serve similar roles in all malignancies. Apart from PPIs, identification, hub genes is one of the important uses of the microarray datasets. The PPI network gene with the highest degree of connectedness is referred to as the hub gene. Due to the hub genes' significant connectivity within the illness network, they are involved in crucial biological processes and have high clinical importance [20]. For the purpose of finding more pertinent biomarkers and important genes for identifying novel primary genes involved in pathogenesis and prognosis, we have also included PPI information.

Typical microarray analysis preprocesses the raw data into normalized data by handling the data value range and missing value. Then, DEGs are calculated to find the statistically significant difference in expression levels of genes. These DEGs are further used for significance, classification, and clustering analysis. Finally, these analysis led to hub gene identification, pathway analysis, and gene function prediction (shown in Fig. 5.6).

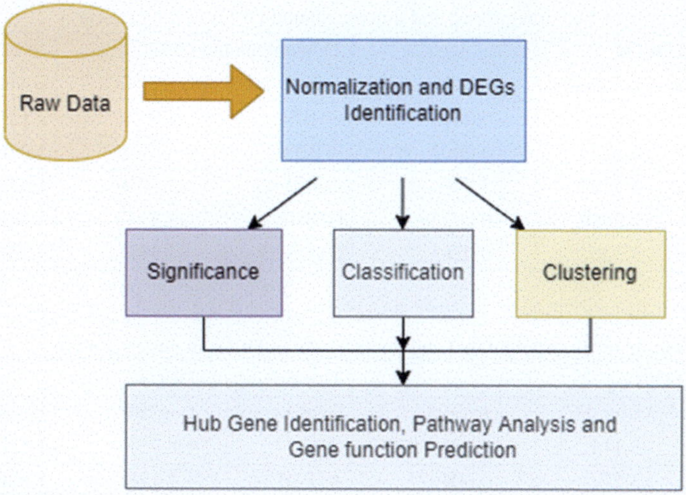

Fig. 5.6 Microarray data analysis

Table 5.2 Characteristics of datasets in this study

GEO Acc. No.	Cancer type	Number of samples	Number of genes
GSE43356	Lung	6	33,297
GSE44024	Breast	4	22,277
GSE45016	Prostate	11	54,675
GSE15459	Gastric	15	56,575

5.5.2.1 Microarray Dataset

Gene expression datasets can be found in the Gene Expression Omnibus (GEO) data repository of the National Center for Biotechnology Information (NCBI), where various genomic data types are submitted by various scientists to summarize their experiments [21]. A small list of different datasets with the number of samples and genes is shown in Table 5.2. Some microarray dataset can also be downloaded from www.biolab.si/supp/bi-cancer/projections/info/.

5.5.2.2 Identification of Differential Gene Expression

When a statistically significant difference in expression levels between two experimental circumstances, such as a disease state and a healthy state, is seen, the differential gene expressions are determined. Finding the differentially expressed genes (DEGs) is crucial to determining which genes are activated or deactivated as a result of a certain disease's invasion of the human body. In this work, we used the R packages edgeR and limma to assess the DEGs between cancer and normal tissues. To identify the significant DEGs, the criteria of $\log 2|FC|$ with a threshold of 1.5 and $p = 0.05$ were utilized. For lung cancer (GSE1987), out of 10,541 genes, 402 are

Table 5.3 Sample of five upregulated and downregulated DEGs in lung cancer

Upregulated DEGs			Downregulated DEGs		
Gene symbol	Log2FC	*p*-value	Gene symbol	Log2FC	*p*-value
COL11A1	4.602	0.0018	FABP4	−4.092	2.12E−06
MMP12	4.1239	0.0002	WIF1	−3.6739	1.07E−05
SPP1	3.9797	5.89E−07	AGER	−3.5039	1.81E−05
TFAP2A	3.9349	0.00339	GDF10	−3.454	1.16E−07
SERPINB5	3.336	0.0285	EXOSC7	−3.313	8.83E−06

Table 5.4 Sample of five upregulated and downregulated DEGs in gastric cancer

Upregulated DEGs			Downregulated DEGs		
Gene symbol	Log2FC	*p*-value	Gene symbol	Log2FC	*p*-value
CST4	5.114	0.001639	GIF	−8.751	1.52E−15
KRT17	3.257	0.004021	ATP4B	−7.126	1.52E−15
VIL1	3.012	4.92E−05	LIPF	−5.544	1.59E−06
ACOT7	2.733	0.003493	LTF	−4.752	1.04E−07
SPP1	2.715	1.11E−07	ATP4A	−4.616	1.00E−14

upregulated and 681 are downregulated. For gastric cancer (GSE2685), out of 4522 genes, 621 are upregulated, while 691 are downregulated. The most statistically significant upregulated and downregulated genes of lung cancer and gastric cancer are listed in Tables 5.3 and 5.4.

5.5.2.3 Construction of PPI Network and Hub Genes Identification

Using the Search Tool for the Retrieval of Interacting Genes STRING (https://stringdb.org/), the proteins corresponding to the up and downregulated genes are used to construct the PPI network. Hub proteins are the higher-degree proteins in the network and are more significant than lower-degree proteins because they interact with a disproportionately large number of other proteins. If these hub proteins are lost, the PPI network may become dispersed into separate clusters. The cytoHubba and MCODE plugin of Cytoscape version 3.6.0 can be used to investigate the whole network and identify the hub genes. A sample list of ten hub genes identified from the gastric cancer PPI network is given in Table 5.5.

5.5.2.4 Survival Analysis of Hub Genes

Using the Kaplan-Meier plotter analytics platform (URL: http://kmplot.com), we assessed the survival curves of each hub gene to further assess the prognostic significance of these genes. The cutoff value is the hazard ratio (HR) with 95% confidence intervals and a log-rank *p* with a 0.05 value. The survival curve shows a decreasing slope as time goes on. A lower survival rate is implied by a steeper slope. It can

Table 5.5 Top ten hub genes of gastric cancer with their degree in PPI network

Gene name	Degree
RPS7	33
UBL4A	30
RPS11	29
RPS27	27
RPL37	24
PSMB2	20
PSMC1	20
RPLP1	18
ADRM1	18
USP14	18

be seen from Fig. 5.7 that the hub genes are significantly correlated with the poor survival of patients with gastric cancer.

5.5.2.5 Classification and Clustering on Microarray Data

Tumor categorization is linked to three primary categories of statistical issues: Unsupervised learning is used to identify novel tumor classes; supervised learning is used to categorize established malignancies; and feature selection is used to identify flag genes that distinguish the various tumor classes.

A good cancer diagnosis and course of therapy depend on a trustworthy and accurate classification of tumors. Human malignancies are now categorized using a number of clinical, morphological, and molecular factors. Diagnostic uncertainty persists despite recent advancements. Additionally, the present classifications are probably heterogeneous and include illnesses that are molecularly separate and have diverse clinical trajectories. Microarray classification is a supervised learning problem that determines a tissue sample's diagnostic categorization based on the phenotype of its expression array. This is accomplished by using labeled data samples and a model that divides fresh data samples into several preset disorders. SVM, nearest-neighbors, random forest classifiers are used in many research for prediction of the class labels of unclassified data. Different clustering methods have been used to find patterns in gene expression data in many literature. The resultant categorization from most cluster analysis techniques is hierarchical. K-means clustering is sometimes used to divide items into multiple clusters without attempting to explain the link between individual objects. There are two separate but connected clustering issues with microarray data, as may be observed. The grouping of the tissues based on the genes (the gene signatures) is one issue, while the clustering of the genes based on the tissues is another (the gene profiles). Cluster analysis frequently uses this duality.

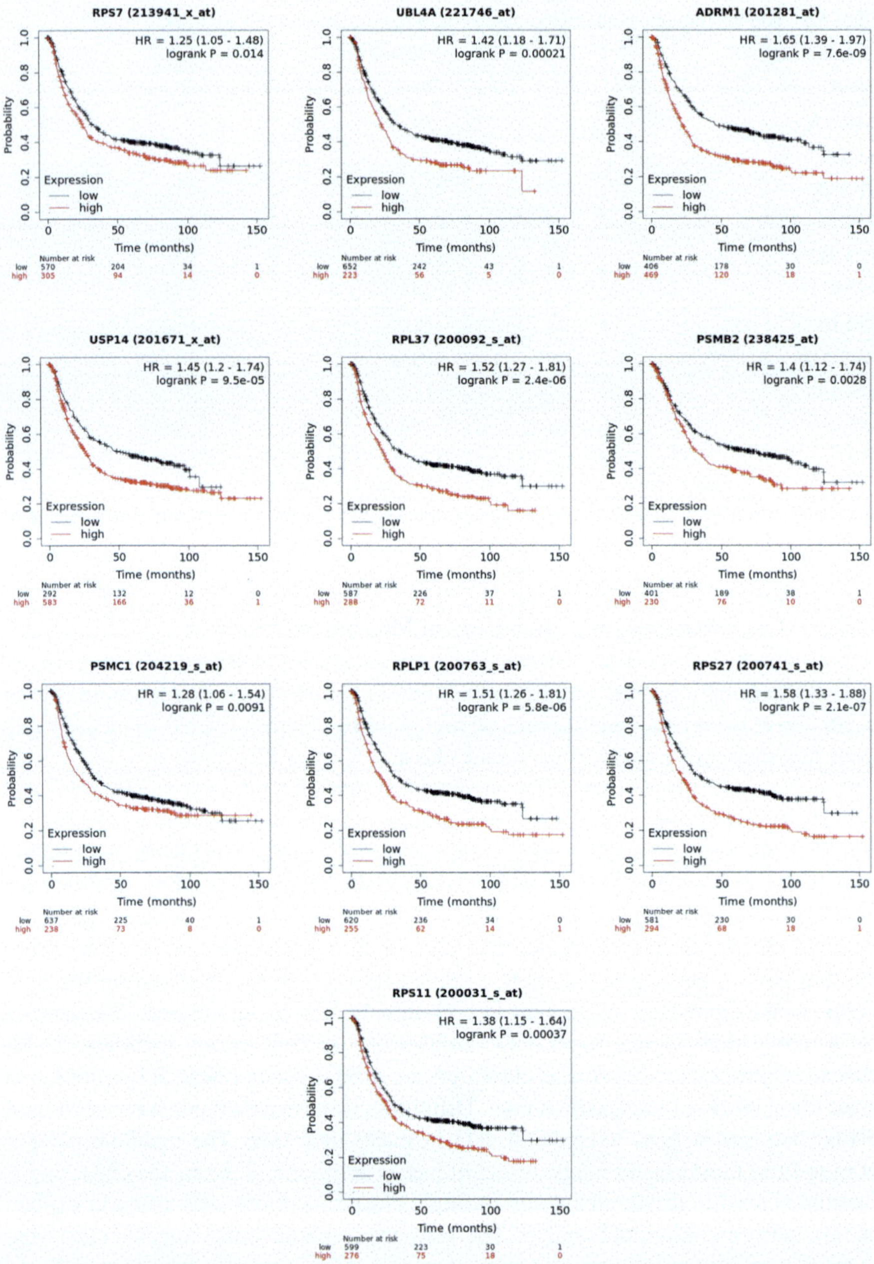

Fig. 5.7 Kaplan-Meier curves depicting the overall survival of hub genes of gastric cancer. HR: hazard ratio

5.5.3 Protein Sequences Data Analysis

Proteins control all biological systems such as regulating all processes, cascading signals, and accelerating some chemical reactions in a cell. A linear sequence of amino acids makes up each protein or peptide. There are 20 amino acids that can be represented by a three or single letter code and described below: Alanine (Ala, A), Arginine (Arg, R), Asparagine (Asn, N), Aspartic acid (Asp, D), Cysteine (Cys, C), Glutamic acid (Glu, E), Glutamine (Gln, Q), Glycine (Gly, G), Histidine (His, H), Isoleucine (Ile, I), Leucine (Leu, L), Lysine (Lys, K), Methionine (Met, M), Phenylalanine (Phe, F), Proline (Pro, P), Serine (Ser, S), Threonine (Thr, T), Tryptophan (Trp, W), Tyrosine (Tyr, Y), Valine (Val, V).

Conventionally, the amino terminal (N) end of a protein's main structure is where it starts, and the carboxyl terminal (C) end is where it ends. It is possible to directly sequence a protein's structure or to deduce it from the DNA sequence. The process of analyzing the amino acid sequence in a protein is known as protein sequencing. Apart from understanding protein structure and function, there are several applications for protein sequencing:

1. Finding the family of proteins that a certain protein belongs to.
2. Predicting the protein's biological location from its target sequence.
3. Prediction of the gene's sequence, which codes for a certain protein.
4. Finding the protein–protein interaction (PPI) in both inter- and intra-species using various computational and experimental techniques.

In this chapter, we review three typical applications of protein sequence data using computational methods: protein structure, protein function, and protein–protein interaction prediction.

5.5.3.1 Protein Structure Prediction from Sequence

One of computational biology's top priorities is the prediction of protein structures, which has applications in both biotechnology and medicine for drug and novel enzyme design. Proteins differ from one another largely in their amino acid composition, which often produces a difference in spatial shape and structure and, as a result, a difference in the biological functions that the protein may perform in cells [22]. Typically, a series of letters spanning a 20-letter alphabet representing the 20 naturally occurring amino acids are used to depict the primary structure or 1D structure. For example, GTREFASEVETTRIALK represents a 1D protein sequence. The chain of amino acids (or residues) that make up a protein folds with alpha helices, beta strands, and non-regular coils in its natural habitat and forms the secondary structure. Based on hydrophobic forces and side chain interactions, these secondary structure is folded again and formed tertiary structure. Lastly, to create protein complexes, many related protein chains can interact or come together. The quaternary structure of proteins is matched by these protein complexes [23]. A 3D protein structure with folds is shown in Fig. 5.8. Protein structure may be ascertained experimentally using methods like X-ray crystallography, NMR spectroscopy, and dual polarization

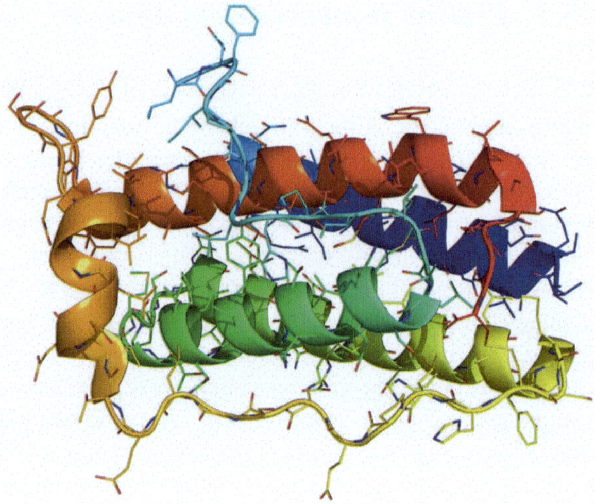

Fig. 5.8 Three-dimensional protein structure

interferometry. The generated protein structures are often uploaded to the Protein Data Bank (PDB), which is a free online resource. Despite advancements in robotics and other fields, it can still be costly, labor-intensive, time-consuming, and often impossible to experimentally determine a protein structure. Although about 40,000 experimentally validated protein structures deposited in PDB to date, large quaternary complexes and membrane protein structure analysis are the most difficult ones due to their complex linkage. Therefore, computational techniques for structure prediction and software have become an important part of cell biology and medicine design.

Protein Structure Prediction Types

The prediction modeling can be divided into two types: template-free and template-based. The template-free method is purely ab-initio methods (de novo methods) that are based on physicochemical properties. These methods demand a tremendous amount of processing power and only use the sequence of amino acids to produce predictions. The amino acid sequence of proteins is also the starting point for template-based modeling, which subsequently discovers proteins with comparable sequences whose structures are known as the homologs. Homologs, or proteins with comparable structures, are frequently connected to one another in terms of evolution. Three-dimensional models that make predictions about the structure of the target protein are created using the discovered homologs as structural templates. According to recent research, when appropriate templates and protein sequence alignments are available, template-based modeling techniques outperform de novo simulations in terms of accuracy and also require less computation power.

Template-Based Protein Structure Prediction

Template-based models specially homology modeling have become more accurate, and with the help of machine learning methods, their range of relevancy has increased.

The model follows the hypothesis that the homologous regions have similar structures. The homology modeling consists of four steps and is summarized in the following steps:

1. Homology modeling needs at least one sequence of known 3D structures with high similarity to the target sequence. This is achieved by using the sequence alignment method. The target sequence is compared with each entry in the sequence database independently either using pairwise sequence alignment or multiple sequence comparison. Popular programs like BLAST, FASTA, PSI-BLAST, and PDB-BLAST are used extensively for sequence similarity searching.
2. Once a list of potential protein templates is obtained, we need to select one or more templates for modeling. The simplest way is to select the structure of the protein as a template whose sequence similarity is highest with the target sequence. Apart from that, the environment (i.e., ligands, solvent, etc.) of both target and potential protein sequences is also to be considered. It may be beneficial to use multiple templates simultaneously by averaging over all of them if it is not clear which is more appropriate.
3. Building a model based on the alignment is the next stage. Some model components might be built simply from known data structures, but there might not be an acceptable one for other components. Therefore, machine learning models like the nearest neighbor method and neural networks are used here for structure prediction specifically in the prediction of loop and side chain conformations.
4. The overall accuracy of comparative models is calculated using the model's energy. It checks the bond lengths and angles of the predicted structure are within normal range or not. It also checks the non-bonded atom distances. These are called internal evaluations and are done by programs like PROCHECK and WHATCHECK. The external evaluation method tests whether a correct template was used or not. This test is important when the sequence alignment is marginal or more than one significant template is available.

A flowchart illustrating these modeling steps is shown in Fig. 5.9.

Softwares for Protein Structure Prediction

A list of software is available for automated structure prediction. This software helps both experts and non-specialists with automated comparative modeling. At the twice-yearly CAFASP and on a regular basis, the LiveBench and EVA evaluate the automated structure prediction algorithms and then validate the outcomes of the software. A list of useful software web servers and their websites link is given in Table 5.6.

5.5.3.2 Protein Function Prediction from Sequence

Protein function is directly linked to its structure, which is indirectly related to its DNA sequence. Basically, protein function is determined by protein structure, which in turn is determined by amino acid sequence, and selection operates on a function

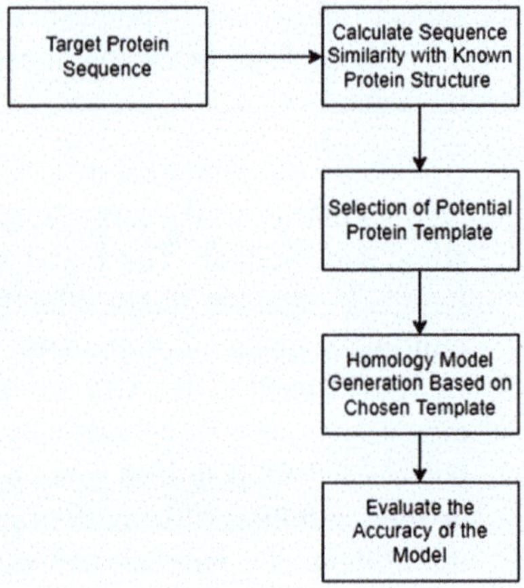

Fig. 5.9 Flowchart for template-based protein structure prediction

Table 5.6 Automated structure prediction software/servers

Name	Description	Link
IntFOLD	Automated web server and some downloadable programs	https://www.reading.ac.uk/bioinf/IntFOLD/
RaptorX	Automated web server and downloadable program	http://raptorx.uchicago.edu/
FoldX	Downloadable program	https://foldxsuite.crg.eu/
HHpred	Interactive web server with help facility	https://arquivo.pt/wayback/20160514083149/http://toolkit.tuebingen.mpg.de/hhpred
MODELLER	Standalone program mainly in Fortran and Python	https://salilab.org/modeller/
CONFOLD	Standalone program mainly in Fortran and Perl	https://github.com/multicom-toolbox/CONFOLD
Molecular Operating Environment (MOE)	Proprietary platform, supported on Windows, Linux, and Mac	https://www.chemcomp.com/

to change allele frequencies in populations [24]. Protein function has different complexity levels including physiological, biochemical, cellular, and phenotypic. With the increase in the discovery of new proteins, the identification of the functionality of these proteins becomes crucial. Therefore, computational methods especially ML techniques are extensively used nowadays to bridge the gap.

In much literature, the taxonomies like a functional catalog (FunCat), enzyme commission (EC), and gene ontology (GO) are used as training datasets for ML algorithms [25]. GO gives an accurate description of the different levels of protein function and is divided into three major types: molecular function (MF), biological process (BP), and cellular component (CC). The term “molecular function” refers to a molecular activity (such as catalysis), and it is frequently anticipated using computational techniques that locate homologs or orthologues. The term "biological process" refers to larger operations carried out by groups of molecular processes, such as a specific metabolic route. The cellular component describes the location(s), such as nucleus or ribosome within a cell where the protein performs its function.

Machine Learning Technique for Protein Function Prediction
The development of effective, affordable sequencing technology, and improvements in computational techniques greatly aid in predicting the yet-to-be-discovered activities of the proteins. In many literatures, protein function prediction has made substantial use of conventional machine learning classifiers including logistic regression, random forests, and support vector machines [26]. They have demonstrated that homology-based function transfer is inferior to integrative prediction techniques. Machine learning methods include the creation and selection of appropriate features of the proteins, as well as algorithms and models that may be taught to carry out this work. The prediction model consists of four steps and is summarized in the following steps:

1. **Features**: Finding appropriate features is a crucial step in every application of machine learning. Using these features, the model may be able to distinguish between two categories of data in a classification problem or fit certain data to an appropriate function in a regression problem. In protein function prediction, the feature vector can be generated from protein features like amino acid sequences, physicochemical properties, and protein–protein interactions. In many applications, gene ontology (GO) terms and (2) enzyme commission (EC) numbers are also used as features.
2. **Training and test set construction**: The non-redundant collection of the feature vectors is divided into training, validation, and test sets, roughly in the proportions of 80/10/10. In supervised learning like support vector machines and random forests, a training dataset is available with a series of output labels corresponding to input vectors. Typically, the training dataset consists of positive and negative datasets. The positive dataset is formed using experimentally verified features with known functions, and the negative dataset is a set of proteins that are known not to carry out a given protein function.
3. **Feature selection**: The use of machine learning techniques is afflicted by the "curse of dimensionality" in many areas, such as biology, where the feature space is so vast that the available data is scarce, which leads to a performance decline. Therefore, it is necessary to exclude some of this abundance of data to arrive at a final set of features that are appropriate for the current issue. Feature selection is the name given to this action. Researchers use different feature selection algorithms like wrapper methods, filter methods, and embedded methods to generate the feature subsets by using the same type of predictive model.
4. **Machine learning algorithms for prediction**: Because the input must be translated into a discrete output, predicting the function of proteins is a classification

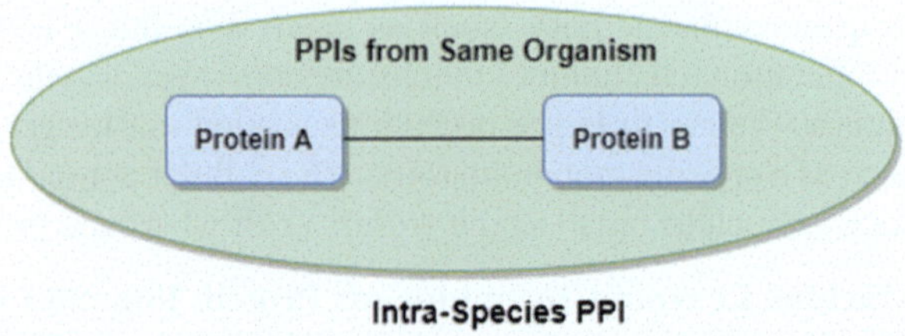

Fig. 5.10 Schematic representation of the intra-species PPI

challenge. In the last several decades, a wide range of machine learning algorithms and models have been created, and they have also been used in numerous settings and applications in protein function prediction. Additionally, deep learning has been used to predict protein function using text-based characteristics. The performance of the algorithms is evaluated using accuracy, precision, recall (sensitivity), specificity, and the F1 score.

5.5.3.3 Protein–Protein Interaction Prediction from Sequence

Although sets of proteins perform their functions independently, a large number of proteins interact with each other to complete the biological activities. Protein–protein interactions (PPIs) occur when two proteins physically bind together to shape functional modules and pathways that complete most of the cellular processes. Therefore, understanding these interactions between proteins is very important to identify the functionalities and disease progression inside a living cell.

PPI Types

PPIs can, in general, be classified into two types. They are intra-species and inter-species. When PPIs happen within the proteins of the same organism, they are called intra-species PPI (Fig. 5.9), whereas inter-species PPI involves proteins of different organisms (Fig. 5.10). Identification of PPIs is crucial for understanding how proteins interact with each other in an organized fashion in a cell to complete cellular functions. Besides formulating drugs, PPIs are also important for protein function analysis, different cellular pathways, and the progression of diseases in the human body. The yeast two-hybrid (Y2H) test, developed by Fields and Song (1989), is the most commonly used cost-effective and scalable in vivo experimental method for detecting interacting protein pairs [12]. Tandem Affinity Purification (TAP) is a mass spectrometry-based approach and also used to identify interacting proteins. It was first successfully used to detect the interactions of the yeast interactome [?]. Besides the Y2H and TAP-tagged method, bacterial two-hybrid test, pull-down assay, and co-localization have been extensively used in many studies to detect the PPIs. Traditional approaches to drug discovery, however, require a lot of resources, take a long period, and frequently produce few drug candidates. Therefore, computational PPI prediction methods are used extensively for faster and more accurate PPI prediction. The majority of earlier studies mainly focused on predicting PPIs within a single creature (intra-species PPI prediction), whereas inter-species PPI prediction has recently attracted increasing attention.

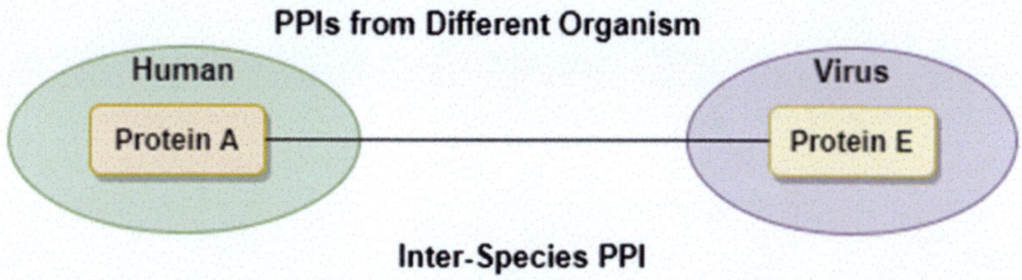

Fig. 5.11 Schematic representation of the inter-species PPI

Protein and Diseases

Diseases can be caused by changes in protein interaction patterns or by the disruption of protein binding. In some cases, mutations change the binding specificity of a protein and create some new and potentially disruptive interactions [3]. For example, a congenital cataract happens when the crystalline proteins are mutated and alter the binding partners [13]. Therefore, PPIs can be used to accumulate information on how their interactions affect the functions of the proteins and cause diseases in the human body. The PPI information can be represented in the form of a graph or network where the nodes represent the proteins and two proteins share an edge if they interact with each other. Figure 5.11 presents a sample PPI network.

PPI Databases

The number of known PPIs has significantly grown during the past several years. Many publicly accessible databases have been created with the goal of gathering and storing information about protein–protein interactions. A list of biological databases that contains experimentally validated PPIs is reported in Table 5.7. The Human Protein Reference Database (HPRD) represents a PPI database of humans that has been generated by manual curation of the published literature by expert biologists. IntAct provides an open-source database used for the storage and analysis of protein interactions of around 960,621 interactions of model organisms. It also contains a web interface to represent textual and graphical representations of protein interactions. The rest of the databases along with the number of interactions and URLs are mentioned in Table 5.7.

Prediction of PPIs

Various types of cancers, cardiovascular diseases (e.g., heart attacks and stroke), chronic respiratory diseases, and diabetes are caused by abnormal mutations of the proteins. Therefore, the prediction of PPI networks can enlighten the molecular basis of a disease, which in turn can help the prevention, diagnosis, and treatment of these diseases.

In [27], Tang et al. developed a new framework called PDMG to predict candidate disease genes/proteins associated with Diabetes mellitus. They achieved quite a large precision on the result compared with Pubmed. An SVM-based predictive model is used for discriminating diabetes versus non-diabetes complexes with an accuracy of 78% in [28]. Zhang et al. analyzed the PPIs for the relation between PM2.5 and diseases like respiratory diseases, cardiovascular diseases, and gastrointestinal diseases

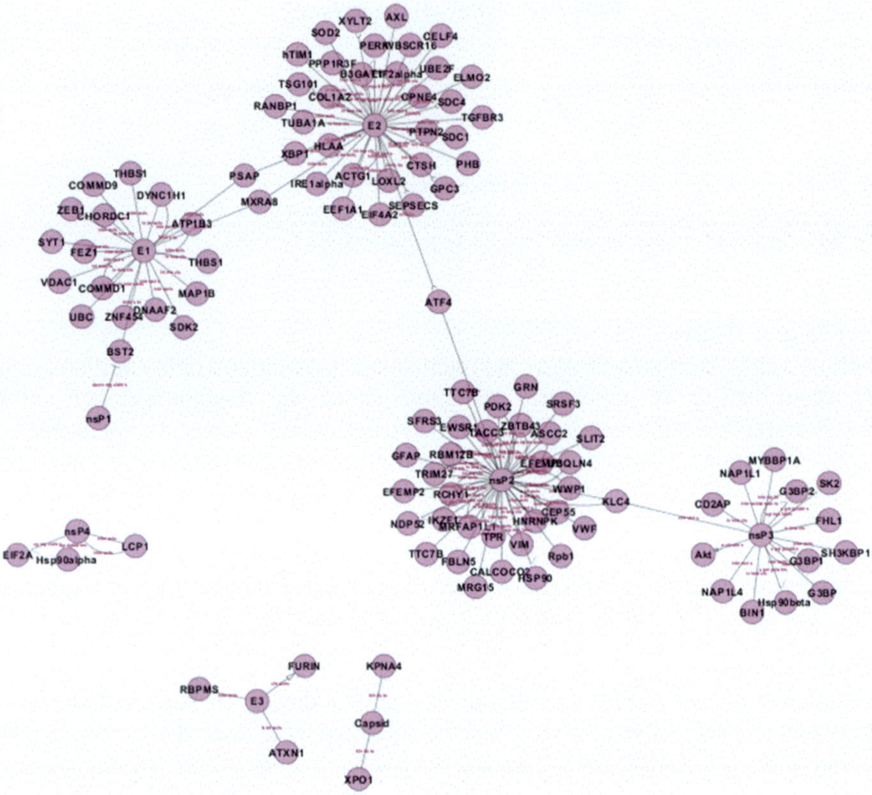

Fig. 5.12 Map of protein–protein interactions network

Table 5.7 Databases of experimental protein–protein interactions

Database	URL	Number of interactions	Organism
BioGRID version 3.5.178	https://thebiogrid.org/	1,740,000	Model organisms
HPRD release 9	http://www.hprd.org/	39,240	Human
IntAct	https://www.ebi.ac.uk/intact/	960,621	Model organisms
BIND	http://bind.ca	198,905	*H. sapiens*, *S. cerevisiae*, *M. musculus*, *H. pylori*
DIP	http://dip.doe-mbi.ucla.edu	73,268	Model organisms
MINT	http://mint.bio.uniroma2.it/mint	131,695	Model organisms

Table 5.8 Computational studies for prediction of various diseases of the human body caused by mutations of the proteins

Disease	Method	References
Diabetes	PDMG method	Tang et al. [27]
	SVM-based prediction	Vyas et al. [28]
	Genome-base study	Lage et al. [31]
Cardiovascular diseases	Analyzed the relation with PM2.5 using PPI database	Zhang et al. [29]
	Similarity measure and clustering methods	Sarajli et al. [32]
	Association score of the proteins using Disease Gene Association (DGA) algorithm	Suratanee et al. [33]
Cancer	Predict cancer proteins using domain–domain interaction (DDI) data, weighted domain frequency score (DFS), and cancer linker degree (CLD)	Huang et al. [34]
	Time-resolved Förster resonance energy transfer technology	Li et al. [35]
	Phenotype specific sub-networks	Engin et al. [30]
Respiratory diseases	Analyzed the relation with PM2.5 using PPI database	Zhang et al. [29]
	Random-walk and network-based closeness approach	Sharma et al. [36]
	Network and matrix-based analysis approach	Garcia et al. [37]

[29]. In [30], the authors used a network- and matrix-based approach to investigate the interaction between the gene, illness, and chemical relationships relevant to respiratory disease. Lage et al. predicted PPI by integrating genome information, and they identified new unexpected genes involved in type 1 diabetes and congenital heart diseases [31]. Table 5.8 highlights the list of methods with references that have been successfully used to predict the PPIs related to cancers, cardiovascular diseases, respiratory diseases, and diabetes in the literature (Fig. 5.12).

Due to the absence of interaction data that has been experimentally confirmed and can be used as a training set, PPI prediction in the host-pathogen network is a difficult problem. Machine learning algorithms are widely used to predict the host-pathogen PPIs. One of the most comprehensive pieces of information for all proteins is their amino acid sequence. As a result, sequence-based approaches along with structure data are widely employed in PPI prediction.

Fu et al. published the HIV1 Human Protein Interaction database which contains a considerable number of Human-HIV PPIs [38]. Doolittle and Gomez predicted humans and HIV PPIs using structural similarity [39]. Apart from HIV, some other PPIs between the virus and host proteins are also studied in different kinds of literature. In [40], Cao et al. proposed a network-based approach "guilt by association" (GBA) to predict PPI between Ebola and humans. The interactions between two different types of viruses, namely the human papillomavirus (HPV) and the hepatitis

Table 5.9 Computational studies for prediction of human-pathogen PPIs

Human-pathogen system	Method	References
Human-HIV1	Bi-clustering-based association rule mining	Mukhopadhyay et al. [45]
	Structural similarity	Doolittle et al. [39]
	Short linear motifs	Becerra et al. [46]
	Protein sequence and network-based features of HIV1 and human and prediction using random forest technique	Tastan et al. [47]
Human-influenza A virus	Two-part controllability analysis	Ackerman et al. [44]
	Structural homology-based approach	Chassey et al. [48]
Human-Salmonella	Sequence similarity-based domain-domain interaction	Schleker et al. [42]
	Structural similarity	Mariano et al. [43]
Human-HCV	SVM-based prediction	Cui et al. [41]
	Graph-based multitask learning-based approach	Kshirsagar et al. [49]
Human-Ebola	Network-based approach "guilt by association" (GBA)	Cao et al. [40]
Human-Dengue	Structural similarity of DENV proteins and human proteins using known interactions	Doolittle et al. [50]
Human-HPV	Protein sequence feature and SVM technique	Cui et al. [41]

C virus (HCV), and human proteins are predicted using an SVM-based technique [41]. Several approaches are also proposed on Salmonella and human PPI prediction [42, 43]. Ackerman et al. applied protein state-related two-part controllability analysis information for PPI prediction between influenza A virus and humans [44]. Table 5.9 outlines the list of techniques that have been effectively used in the literature to anticipate human-pathogen interactions.

5.6 Conclusion

The recent rapid advancements in hardware technology have given biologists new opportunities to gather data in a variety of application areas, including genomics, protein modeling, sequence analysis, etc. The development of technologies has also presented enormous challenges. The chapter focuses on machine learning algorithm development and novel data mining analysis for biological data. When examining DNA sequences, a huge variety of algorithms can perform effectively, but their mining outputs are very sensitive and particular, which will cause a significant variation when used. Therefore, the issue meriting everyone's attention and research is how to include the biological importance of DNA sequences in the data mining procedure. Protein structure and function prediction is a promising field of computational biology. Although the prediction accuracy has not crossed 80%, we still hope for a better prediction algorithm in the future. DNA microarray technology and its analytical methods have been highly evolved to diagnose and identify the genes that cause genetic disorders especially cancer and cystic fibrosis. PPI prediction for inter- and intra-species networks generates a large set of potential interactions. These findings may influence the understanding of different disease progression in the human body at the molecular level and accelerate new strategies for vaccination and treatment of viral infection. Finally, as a word of caution, all the results produced by these computational methods should be taken carefully. Although the prediction methods are very useful for speeding up gene discovery and handling large databases, biological expertise is required to confirm the results.

References

1. Gauthier J, Vincent AT, Charette SJ, Derome N (2019) A brief history of bioinformatics. Brief Bioinform 20(6):1981–1996
2. Yang A, Zhang W, Wang J, Yang K, Han Y, Zhang L (2020) Review on the application of machine learning algorithms in the sequence data mining of DNA. Front Bioeng Biotechnol 8:1032
3. Aerssens J, Armstrong M, Gilissen R, Cohen N (2001) The human genome: an introduction. Oncologist 100–109
4. Dixit P, Prajapati GI (2015) Machine learning in bioinformatics: a novel approach for dna sequencing. In: 2015 fifth international conference on advanced computing & communication technologies. IEEE, pp 41–47
5. Rashed AEED, Amer HM, El-Seddek M, Moustafa HED (2021) Sequence alignment using machine learning-based Needleman-Wunsch algorithm. IEEE Access 9:109522–109535
6. Gunasekaran H, Ramalakshmi K, Rex Macedo Arokiaraj A, Deepa Kanmani S, Venkatesan C, Suresh Gnana Dhas C (2021) Analysis of DNA sequence classification using CNN and hybrid models. Comput Math Methods Med 2021
7. Remita AM, Diallo AB (2019) Statistical linear models in virus genomic alignment-free classification: application to hepatitis C viruses. In: 2019 IEEE international conference on bioinformatics and biomedicine (BIBM), San Diego, CA, USA, Nov 2019
8. Millán Arias P, Alipour F, Hill KA, Kari L (2022) DeLUCS: deep learning for unsupervised clustering of DNA sequences. PLoS ONE 17(1):e0261531
9. Rashed AEED, Obaya M, El-Din Moustafa H (2021) Accelerating DNA pairwise sequence alignment using FPGA and a customized convolutional neural network. Comput Electr Eng 92:107112
10. Zhao Y, Ma P, Lan J, Liang C, Ji G (2008) An improved ant colony algorithm for DNA sequence alignment. In: 2008 international symposium on information science and engineering, vol 2. IEEE, pp 683–688

11. Huo HW, Xiao ZW (2007) A multiple alignment approach for DNA sequences based on the maximum weighted path algorithms. Ruan Jian Xue Bao (J Softw) 18:185–195. https://doi.org/10.1360/jos180185
12. Nguyen NG, Tran VA, Phan D, Lumbanraja FR, Faisal MR, Abapihi B, Satou K (2016) DNA sequence classification by convolutional neural network. J Biomed Sci Eng 9(5):280–286
13. James BT, Luczak BB, Girgis HZ (2018) MeShClust: an intelligent tool for clustering DNA sequences. Nucl Acids Res 46(14):e83
14. Chappell T, Geva S, Hogan J (2017) K-means clustering of biological sequences. In: Proceedings of the 22nd Australasian document computing symposium, pp 1–4
15. Guo T, Ma H, Zhou Y (2019) Bioinformatics analysis of microarray data to identify the candidate biomarkers of lung adenocarcinoma. PeerJ 7:e7313
16. Wekesa JS, Luan Y, Meng J (2020) Predicting protein functions based on differential coexpression and neighborhood analysis. J Comput Biol
17. Karimizadeh E, Sharifi-Zarchi A, Nikaein H, Salehi S, Salamatian B, Elmi N, Gharibdoost F, Mahmoudi M (2019) Analysis of gene expression profiles and protein-protein interaction networks in multiple tissues of systemic sclerosis. BMC Med Genomics 12(1):1–12
18. Wu C, Zhu J, Zhang X (2012) Integrating gene expression and protein-protein interaction network to prioritize cancer-associated genes. BMC Bioinform 13(1):182
19. Guda P, Chittur SV, Guda C (2009) Comparative analysis of protein-protein interactions in cancer-associated genes. Genomics Proteomics Bioinform 7(1–2):25–36
20. Xiong Y, You W, Wang R, Peng L, Fu Z (2017) Prediction and validation of hub genes associated with colorectal cancer by integrating PPI network and gene expression data, Hindawi. BioMed Res Int 2017:ID 2421459. https://doi.org/10.1155/2017/2421459
21. Barrett T, Edgar R (2006) Mining microarray data at NCBI's gene expression omnibus (GEO). Methods Mol Biol 338:175–190
22. Al-Lazikani B, Jung J, Xiang Z, Honig B (2001) Protein structure prediction. Curr Opin Chem Biol 5(1):51–56
23. Cheng J, Tegge AN, Baldi P (2008) Machine learning methods for protein structure prediction. IEEE Rev Biomed Eng 1:41–49
24. Whisstock JC, Lesk AM (2003) Prediction of protein function from protein sequence and structure. Q Rev Biophys 36(3):307–340
25. Bonetta R, Valentino G (2020) Machine learning techniques for protein function prediction. Proteins: Struct Funct Bioinform 88(3):397–413
26. Gligorijević V, Renfrew PD, Kosciolek T, Leman JK, Berenberg D, Vatanen T, Bonneau R (2021) Structure-based protein function prediction using graph convolutional networks. Nat Commun 12(1):1–14
27. Tang X, Hu X, Yang X, Fan Y, Li Y, Hu W, Liao Y, Cai M, Zheng W, Peng W, Gao L (2016) Predicting diabetes mellitus genes via protein-protein interaction and protein subcellular localization information. BMC Genomics 17(4):433
28. Vyas R, Bapat S, Jain E, Karthikeyan M, Tambe S, Kulkarni BD (2016) Building and analysis of protein-protein interactions related to diabetes mellitus using support vector machine, biomedical text mining and network analysis. Comput Biol Chem 65:37–44
29. Zhang Q, Zhang PW, Cai Y-D (2016) The use of protein-protein interactions for the analysis of the associations between PM2.5 and some diseases. BioMed Res Int 2016
30. Engin HB, Guney E, Keskin O, Oliva B, Gursoy A (2013) Integrating structure to protein-protein interaction networks that drive metastasis to brain and lung in breast cancer. PLoS ONE 8(11):e81035
31. Zhang Q, Zhang P-W, Cai Y-D (2016) The use of protein-protein interactions for the analysis of the associations between PM2.5 and some diseases. BioMed Res Int 2016
32. Sarajlic A, Janji V, Stojkovi N, Radak D, Przulj N (2013) Network topology reveals key cardiovascular disease genes. PLoS ONE 8(8):e71537
33. Suratanee A, Plaimas K (2018) Network-based association analysis to infer new disease-gene relationships using large-scale protein interactions. PLoS ONE 13(6):e0199435

34. Huang C-H, Peng H-S, Ng K-L (2015) Prediction of cancer proteins by integrating protein interaction, domain frequency, and domain interaction data using machine learning algorithms. BioMed Res Int 2015
35. Li Z, Ivanov AA, Su R, Gonzalez-Pecchi V, Qi Q, Liu S, Webber P, McMillan E, Rusnak L, Pham C et al (2017) The OncoPPi network of cancer-focused protein-protein interactions to inform biological insights and therapeutic strategies. Nat Commun 8:14356
36. Sharma A, Kitsak M, Cho MH, Ameli A, Zhou X, Jiang Z, Crapo JD, Beaty TH, Menche J, Bakke PS et al (2018) Integration of molecular interactome and targeted interaction analysis to identify a COPD disease network module. Sci Rep 8(1):14439
37. Garcia B, Datta G, Cosgrove GP, Strong M (2014) Network and matrix analysis of the respiratory disease interactome. BMC Syst Biol 8(1):34
38. Ako-Adjei D, Fu W, Wallin C, Katz KS, Song G, Darji D, Brister JR, Ptak RG, Pruitt KD (2014) HIV-1, human interaction database: current status and new features. Nucleic Acids Res 43(D1):D566–D570
39. Doolittle JM, Gomez SM (2010) Structural similarity-based predictions of protein interactions between HIV-1 and homo sapiens. Virol J 7(1):82
40. Cao H, Zhang Y, Zhao J, Zhu L, Wang Y, Li J, Feng Y-M, Zhang N (2017) Prediction of the Ebola virus infection related human genes using protein-protein interaction network. Comb Chem High Throughput Screen 20(7):638–646
41. Cui G, Fang C, Han K (2012) Prediction of protein-protein interactions between viruses and human by an SVM model. BMC Bioinform 13:S5
42. Schleker S, Garcia-Garcia J, Klein-Seetharaman J, Oliva B (2012) Prediction and comparison of salmonella and human and salmonella and Arabidopsis interactomes. Chem Biodivers 9(5):991–1018
43. Mariano R, Wuchty S (2017) Structure-based prediction of host-pathogen protein interactions. Curr Opin Struct Biol 44:119–124
44. Ackerman EE, Alcorn JF, Hase T, Shoemaker JE (2019) A dual controllability analysis of influenza virus-host protein-protein interaction networks for antiviral drug target discovery. BMC Bioinform 20(1):297
45. Mukhopadhyay A, Maulik U, Bandyopadhyay S (2012) A novel biclustering approach to association rule mining for predicting HIV-1-human protein interactions. PLoS ONE 7(4):e32289
46. Becerra A, Bucheli VA, Moreno PA (2017) Prediction of virus-host protein-protein interactions mediated by short linear motifs. BMC Bioinform 18(1):163
47. Tastan O, Qi Y, Carbonell JG, Klein-Seetharaman J (2009) Prediction of interactions between HIV-1 and human proteins by information integration. In: Biocomputing. World Scientific, pp 516–527
48. De Chassey B, Meyniel-Schicklin L, Aublin-Gex A, Navratil V, Chantier T, Andre P, Lotteau V (2013) Structure homology and interaction redundancy for discovering virus-host protein interactions. EMBO Rep 14(10):938–944
49. Kshirsagar M, Carbonell J, Seetharaman JK (2013) Multitask learning for host-pathogen protein interactions. Bioinformatics 29(13):i217–i226
50. Doolittle JM, Gomez SM (2011) Mapping protein interactions between dengue virus and its human and insect hosts. PLoS Negl Trop Dis 5(2):e954

Chapter 6
Data Classification Through Cognitive Computing

6.1 Introduction

A brain–computer interface (BCI) is a direct link between the human brain and an external system. Computer–brain interfaces are used to send sensory information from the brain or to activate the brain with electrical signals created artificially. BCI offers a wide range of applications, including wheelchair control, spelling aids, robot control, and cursor control, among others. Cursor control is one of the most significant applications for impaired individuals to communicate successfully with the application interface. Cursor movement is classified as either single target or multitarget. Electroencephalography (EEG), functional magneto resonance imaging (fMRI), near-infrared spectroscopy (NIRS), and magneto encephalography are all methods for capturing brain signals (MEG). EEG is the most commonly employed among them for signal collection. Based on frequency, EEG divides the brain rhythm into five categories (α, β, θ, δ, γ). People can efficiently operate the cursor based on different types of brain rhythms by converting the scale voltage of the brain signal into amplitude level as a cursor. However, because disabled people or people with neuromuscular disorders are unable to use their left or right limbs, they require a BCI tool that can operate the cursor. If impaired people can properly move the cursor, they can reply to the outside world relatively fast. For multitarget-based BCI, disabled persons can use a machine learning technique to select a series of nearby targets and then control the cursor movement to each target one by one using an efficient method. As paralyzed persons are unable to convey their emotions owing to a significant problem, emotion detection for them is a promising potential in cognitive research. Emotion may be identified from the brain signal because it is collected in several electrodes, each of which is linked to underlying lobes in the brain, such as electrode F3, P3, which is put over the frontal and parietal lobes, respectively. However, the activation or deactivation of brain signals in each lobe of the brain is linked to a certain emotion, which can be classified using a sophisticated machine learning algorithm. Disabled people can interact effectively with the outside world if their emotions are properly categorized.

S. Chakraborty and L. Dey, *Computing for Data Analysis: Theory and Practices*, Data-Intensive Research, https://doi.org/10.1007/978-981-19-8004-6_6

6.2 Background

The cursor position is affected by the voltage level of the brain signal on the scalp. The paper [1] describes an efficient method of cursor navigation to the target. However, using various categorization and signal processing techniques, various brain activity patterns of impaired persons are detected and converted into essential directives. BCI offers a wide range of applications, including wheelchair control, spelling devices, and robot control, among others. Based on electrode placement in the brain, BCI has three techniques (invasive, partially invasive, and non-invasive). In this chapter, we will look at the EEG signal from the non-invasive device. Different electrodes record electric brain impulses, which are then classified using EEG. The signal's useful features are then extracted, and then applied classification accordingly. Therefore, the signal is transformed into intended actions for disabled people. Finally, the signal is sent to an application interface, where it is transformed into appropriate action for people with disabilities. Due to the rapidly expanding interest in EEG-based BCI, several studies have been conducted on the feature categorization process in BCI. The overall BCI architecture is represented in Fig. 6.1 [2].

6.2.1 Basic Components of BCI

Signal capture, signal processing, feature extraction and classification, and application interface are the four sections of BCI. Each section's specifics have been detailed below:

(A) **Acquisition of Signal**

The dendritic synaptic stimulation produces current that is measured by EEG signal. The current generates an electric field on the scalp, which is then distributed throughout the body [3]. During signal acquisition, internal (brain-generated) or external (scalp-generated) noise is generated. The noisy signal

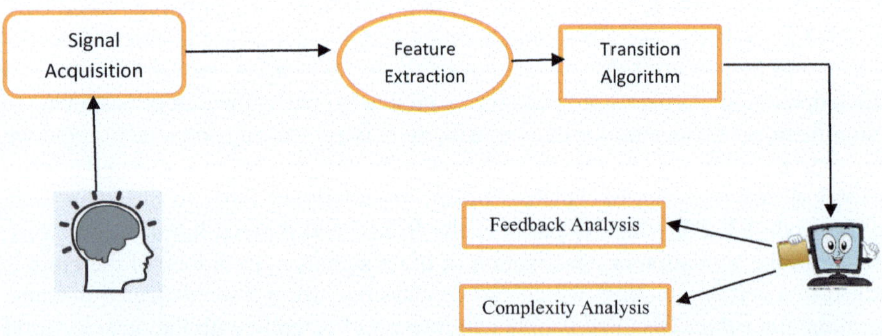

Fig. 6.1 BCI system architecture

may diminish the signal's strength; thus, it is sent to the next stage of BCI (signal processing) to improve the signal's strength.

Classification of EEG Signal

1. Alpha Rhythm: This term refers to a state of physical and mental repose. It features a low voltage (77) and a high frequency (8–13 Hz).
2. Beta Rhythm: Use a high frequency (13–20 Hz) to describe the state of consciousness [4].
3. Theta Rhythm: Use low frequency (4–8 Hz) to describe deep and regular sleep [4].
4. Delta Rhythm: Use low frequency (0–4 Hz) to describe pathological situations of neuronal difficulty (comma) [4].
5. Gamma Rhythm: This is a brainwave that is frequently connected with high-frequency activity (21–30 Hz) [4].

Classification of Signals

1. P300: This signal is generated inside the ERP system. After a specific external or internal event happens, an ERP response is created [5].
2. SSVEP: In response to a visual input, this signal is created.
3. Sensorimotor Rhythm (SMR): It occurs within 8–12 Hz frequency range [6]. It can identify BCI's recent and future therapeutic uses potential [7].
4. Slow Cortical Potential (SCP): The change in dendritic depolarization stage influences it. The sum of synchronized potential created in a specific time range is referred to as negative SCP [8].

(B) **Signal Processing**

The noise in the brain signal must be eliminated in order to recognize and classify features. The signal-to-noise ratio (SNR) is a measurement of the power of both the signal and the noise. A high signal-to-noise ratio (SNR) indicates that signal corruption is minimal. Background neurological activity generates noise, as artifacts created by biological and external sources [9]. Therefore, the translation algorithm is fed by the denoised actual signal.

(C) **Feature Extraction and Classification**

The relevant features connected to the BCI must be extracted after the signal strength has been improved utilizing the signal processing phase. During the feature extraction phase, features are extracted. For extracting meaningful features from the signal, many approaches such as autoregressor, independent component analysis (ICA), common spatial pattern (CSP), continuous wavelet transform (CWT), and others have been utilized. After extracting the features, it is time to look for features that are relevant to BCI applications. BCI input typically handles a large number of attributes. The feature subset selection procedure aids researchers in identifying key feature subsets extracted using various feature extraction methods. After the useful characteristics have been extracted, the classification procedure has been applied to those features.

(D) **Application Interface**

The output of classification has been passed to the final step of the BCI after the feature classification phase (i.e., application interface). The application interface's translation method transforms the classification phase's output into BCI device-specific commands. This device-specific command provides feedback to people, allowing them to interact with the outside world.

6.2.2 Feature Extraction

Some essential qualities of these features are detailed below in terms of the BCI system's design [10].

1. ***Outliers***: BCI signals are embedded with outliers as EEG signal has low SNR.
2. ***Dimensionality***: BCI consists of high-dimensional feature vectors, and numerous features from the brain are extracted throughout time and integrated into a single vector.
3. ***Information About Time***: Because the human brain's signal pattern changes over time, BCI features must include temporal information.
4. ***Small Training Sets***: BCI requires a least amount of computational time by the use of a small training dataset.

Feature Extraction Using Fourier Transform
The time and frequency levels of the EEG signal have been categorized. Unwanted noise, on the other hand, takes use of the signal properties. The valuable features are taken out from the EEG signal through fast Fourier transform (FFT) [11, 12]. In the paper [11], some interesting feature creation approaches are explored.

Feature Extraction Model
Table 6.1 describes the outline of various feature selection methods.

6.2.3 Feature Classification

There are two elements to the feature classification method. (1) Regression and (2) classification, but in recent years, the classification strategy has proven to be the most effective [26]. When classification algorithms are used for categorical inputs, regression methods are effective for quantitative data. As "binary classification" is frequently called, classification algorithms are usually employed for two-class issues. As a result, for two-target applications, the classification strategy may be more useful. Regression algorithms are superior for multiple targets, according to the author of research [27]. Table 6.2 provides a detailed overview of numerous feature classification techniques.

Table 6.1 Outline of feature selection methods

Category	Method	Properties	Applications
Dimension reduction	Principal component analysis (PCA)	(1) Transformation in linear space	[13, 14]
		(2) Transformation of correlated observations into a set of variables which are not correlated	
		(3) Important for outliers and dimension reduction method	
	Independent component analysis (ICA)	(1) Reduce the time it takes for signals to be processed and enhance the speed at which data is communicated	[15]
		(2) Decompose EEG signals into separate, independent components, which is useful for removing artifacts	[16, 17]
Space	Common spatial pattern (CSP)	(1) Maximizes the difference between the variances of two classes' data density distributions	[18]
		(2) Asynchronous BCIs perform better than synchronous BCIs; however, asynchronous BCIs perform poorly	[19]
Time–frequency	Autoregressive component (AR)	(1) It can be used for a variety of purposes, including data compression and classification	[20]
		(2) EEG signal is represented in a concise and efficient manner	[21]
	Continuous wavelet transform (CWT)	(1) Signals relating to events in the brain are useful	[2]
		(2) The signal's frequency and temporal information are both delivered	[22]
	Discrete wavelet transform (DWT)	(1) CWT's redundancy and complexity are reduced	[23]
		(2) Classify the EEG signal by dividing it into overlapping windows	[24]

(continued)

Table 6.1 (continued)

Category	Method	Properties	Applications
		(3) From multiscale data, DWT and the modified fractal dimension approach extract features	[25]

Combination of Classifiers: Ensemble Methods

Ensemble methods are a collection of learning classifiers and then use a (weighted) vote of their predictions to categorize fresh data points. Bagging, boosting, and random forest are three approaches used in ensemble methods.

Bagging

First, imbalance issue of data is removed from EEG dataset and then a combination of bagging with SVM is used to enhance the performance of BCI-P300, according to the author of paper [40]. This strategy minimizes the training variance at training phase and improves the prediction accuracy at validation phase [41].

Boosting

Gradient boosting is utilized to identify EEG-based P300 BCI applications. The gradient boosting is a quite good ensemble model in terms of accuracy and complexity [42]. However, noise can reduce the performance factor of this model [43].

Random Forest

Individual classifiers and independent trees are constructed using a random selection of feature subspace in this method [44]. To arrive at the final forecast, the output from many classifiers was combined using the majority vote [45]. The author of paper [46] describes an efficient word typing P300 system that uses a random forest classifier to increase classification accuracy by mixing numerous decision trees.

6.3 Result Analysis

A. ***Comparative Analysis Between Basic Classifiers***

- ***Analysis Among FLDA, BLDA, NN***

  In a P300-based BCI system, Fig. 6.2 shows that BLDA has a greater accuracy than the other two classifiers (FLDA, NN). FLDA suffers from an overfitting problem (n) due to the feature number that is greater than the input numbers. Because of its inherent model complexity, the neural network (NN) produces the worst results. Figure 6.3 shows that the BLDA has less error rates than

Table 6.2 Outline of various classification techniques

Category	Method	Characteristics	Utilizations
Generative model	Bayesian probabilistic analysis	(1) For finding the posterior output, use previous statistical information of the feature and a parameter estimate technique (maximum likelihood)	[28]
Linear	Linear discriminant analysis (LDA)	(1) To split features into distinct classes depending on discriminant function, draw a hyperplane	[5]
		(2) When there is a lot of noise and a lot of dimensionality, it fails	[29]
		(3) It can be used to classify brain event related potential (ERP) signals	[30]
		(4) Bayesian linear discriminant analysis (BLDA) and Fisher linear discriminant analysis are improved variants of LDA (FLDA)	[30]
	Support vector machine (SVM)	(1) Maximizes the distance between the separating hyperplanes and the nearest training data points	
		(2) The optimal number of channels is determined by a constant	[31]
		(3) It can be used in both linear and nonlinear approaches	[32]
		(4) In the presence of outliers, it produces poor results, necessitating regularization	[33]
Nonlinear	K-nearest neighbor (KNN)	(1) Uses distance measure functions of the validation data and their neighbors	[34]
		(2) When the number of feature vectors is minimal and locally distributed, it produces good results	[35]

(continued)

Table 6.2 (continued)

Category	Method	Characteristics	Utilizations
		(3) Gives high performance in pattern recognition problems	[6, 21]
	Artificial neural network (ANN)	(1) Human brain functionality influences this neural network model	
		(2) Unbalanced class distribution is not recommended	[7]
		(3) Provides high accuracy categorization for a variety of mental and physical tasks	[36]
	Quadratic discriminant	(1) For each class, a distinct covariance matrix is used	
	analysis (QDA)	(2) Left and right limb movements are controlled by this method in BCI	[37]
Graph-inspired methods	Hidden Markov model (HMM)	(1) A dynamic Bayesian network	
		(2) Useful for classifying EEG data in real time while imagining a left or right-hand movement	[38]
		(3) For finger motions, classify the ERD and ERS aspects of the EEG signal	[39]

the other classifiers for a high number of data points due to the removal of the over fitting problem.

- ***Comparison Among LDA, ANN, SVM***

  The paper [29] evaluated linear (LDA) and nonlinear (SVM, NN) classifiers for classification of EEG signals for all different emotions. However, every exercise (relaxing, writing a letter, solving puzzles, counting, and spinning items) required a few seconds of brain signaling, and each task was repeated five times.

  Figures 6.4 and 6.5 provide a comparison of linear (LDA) and nonlinear (NN, SVM) classifiers for various emotional operations. Figure 6.4 deduces that SVM provides better classification results than LDA. For the EEG data of high dimension, nonlinear classifiers, particularly SVM, outperform LDA.

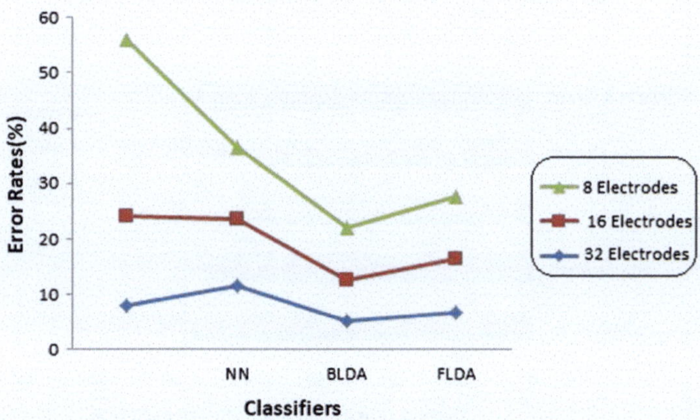

Fig. 6.2 Relation between error rates of classifiers with respect to number of electrodes

Fig. 6.3 Performance measure of NN, FLDA, BLDA

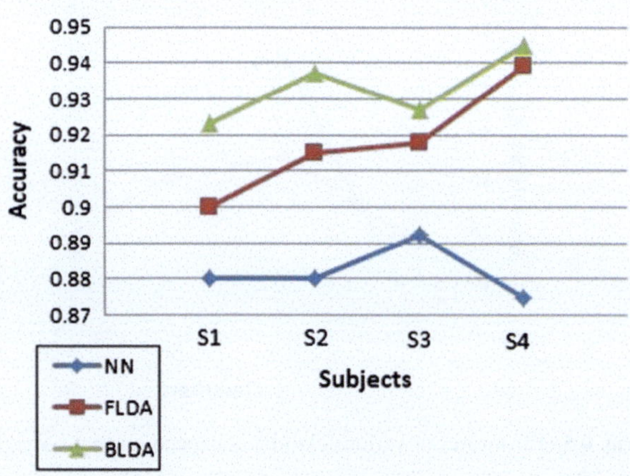

In a multilayer backpropagation network, NN has trouble finding the right amount of hidden units; therefore, it cannot provide a good generalization model. Because SVM may select a small and efficient subset of characteristics, the number of needed electrodes is reduced, reducing noise and improving accuracy for all mental tasks [29]. In Fig. 6.5, every trial's window size determines how different mental tasks are classified with respect to the time. However, SVM with genetic algorithm (GA) has lower error rates than other classifiers.

- ***Comparative Analysis Among QDA, LDA, KNN***

The EEG signal is initially processed, and then useful properties (wavelet coefficient, PSD, average band power, and so on) are retrieved [37]. The features are then input into different classifiers one by one. The various classifiers (LDA, QDA, *K*NN) can categorize the motion of left and right limb from the filtered EEG signal after the feature extraction phase.

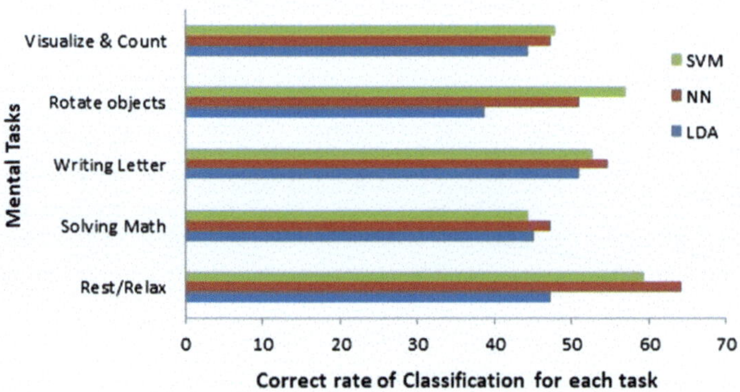

Fig. 6.4 Accuracy analysis between LDA, NN, SVM based on different emotions

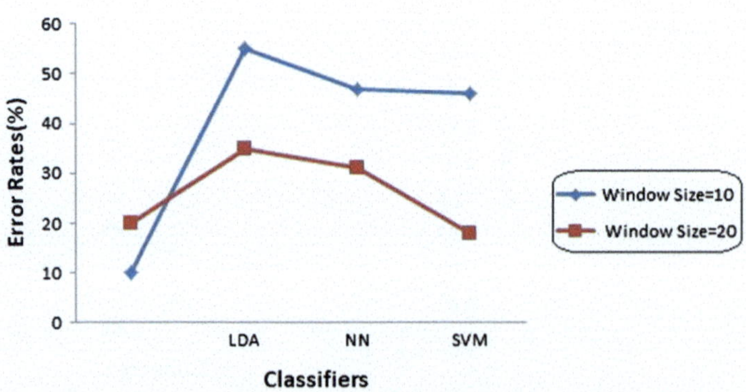

Fig. 6.5 Error rates of various classifiers depend on trial-based window size

From Figs. 6.6, 6.7, and 6.8, wavelet coefficient with QDA provides better result and both *K*NN and QDA provide identical accuracy. When average band power is set as a characteristic, however, *K*NN outperforms all others. We evaluate the error rate of each classifier for the three features in Fig. 6.9 and find that *K*NN and LDA classifiers have better error rates for wavelet coefficient than QDA, while LDA and QDA have larger error rates for rest of the features.

- ***Comparative Analysis Between SVM and HMM***

The categorization of NIRS signals in BCI using SVM and HMM to drive a word speller application is discussed in paper [47]. The brain's NIRS signal is recorded here, and the useful features for motor imaging tasks are extracted. After that, classifiers (HMM, SVM) that suit the trends of retrieved features are used to classify the signal. The classifier strives to provide the best classification result possible so that the user can move the pointer around the word speller application and select a box by choosing between left and right-hand images.

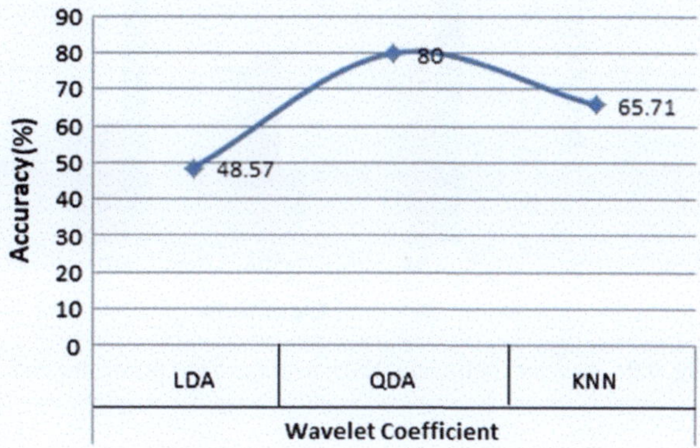

Fig. 6.6 Wavelet-based accuracy analysis among QDA, LDA, *K*NN

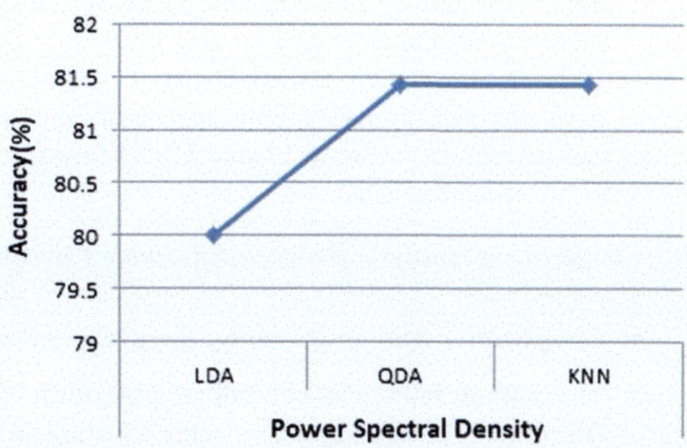

Fig. 6.7 Accuracy comparison among QDA, LDA, *K*NN with PSD coefficient

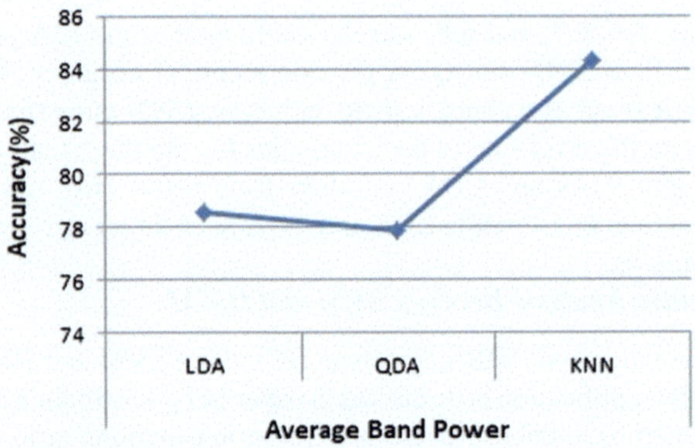

Fig. 6.8 Comparison on accuracy among LDA, QDA, *K*NN with average band power coefficient

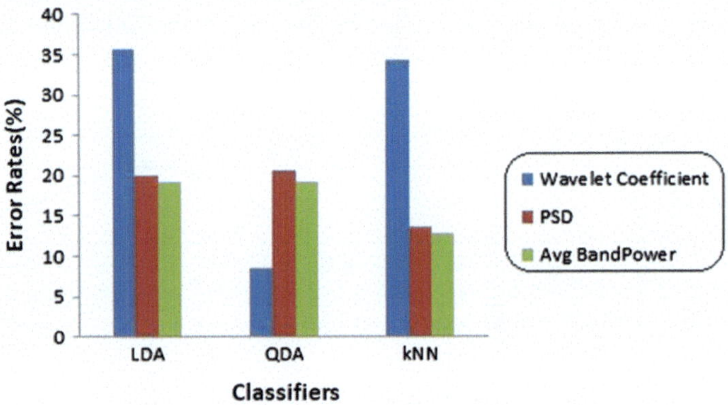

Fig. 6.9 Wavelet coefficient, PSD, average band power-based error rates comparison

In motor imagery operations, the HMM outperforms the SVM, as shown in Figs. 6.10 and 6.11. HMM and SVM were used in the experiment for each pattern (left and right-hand motor imagery). The HMM model outperforms SVM in all dynamic tasks because it is based on a probabilistic model and performs well in the dynamic nature of hemodynamic time series data. We can deduce from Fig. 6.11 that HMM has a lower error rate than SVM for motor imagery tasks.

B. ***Comparative Analysis Between Ensemble Classifiers***

- ***Comparative Among Boosting, Bagging, and Ensemble Voting Techniques***

The author built a novel voting algorithm and compared it to bagging (Breiman's CART algorithm) and boosting in his paper [48, 49] (Quinlan

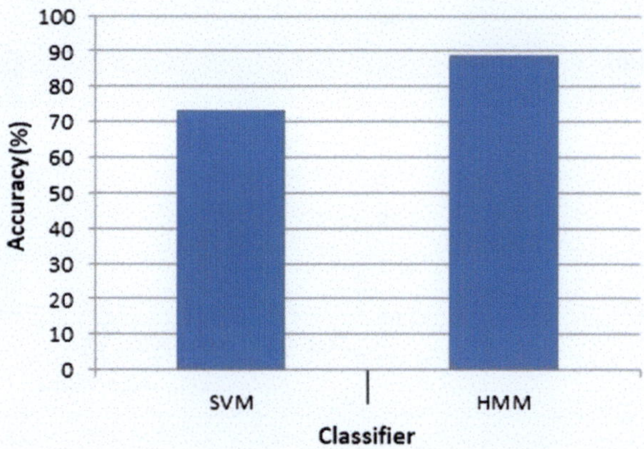

Fig. 6.10 Comparison of accuracy with respect to motor imagery tasks

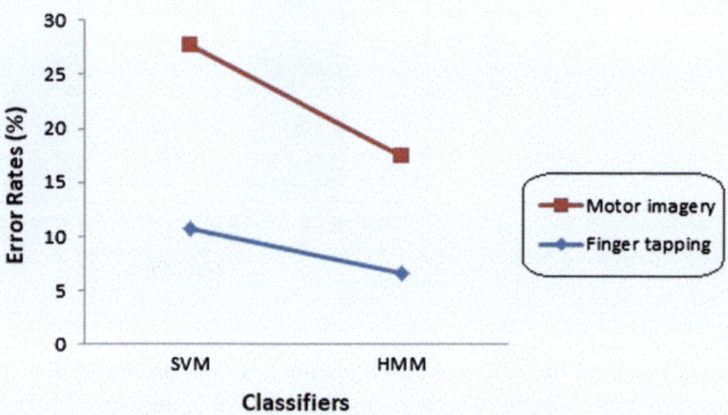

Fig. 6.11 Comparison of error rates with respect to motor imagery tasks

C4.5 algorithm). This study's author divides the error into two categories (bias and variance) and seeks to reduce these two terms utilizing voting features (pruning the massive tree, probabilistic estimation of the feature's coefficients, and so on). We plotted the failure rates and accuracy evaluation for voting algorithms using boosting and bagging in Figs. 6.12 and 6.13.

Because it minimizes failure rates and boosts performance gain, the voting technique is more accurate than the bagging (CART) and boosting (C4.5) algorithms, as shown in the figures (Figs. 6.12 and 6.13). After incorporating

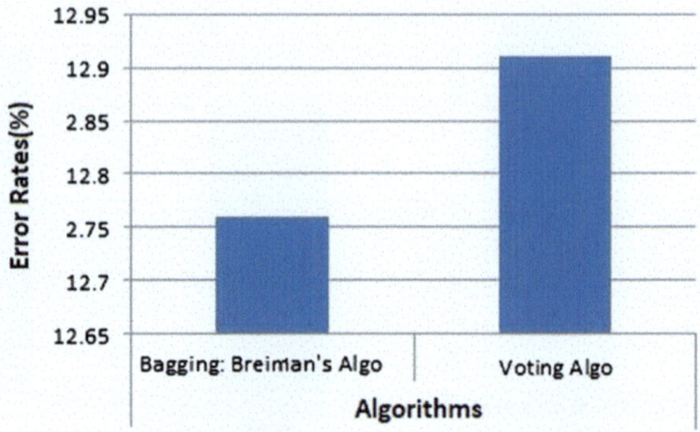

Fig. 6.12 CART versus voting algorithm with respect to error rates

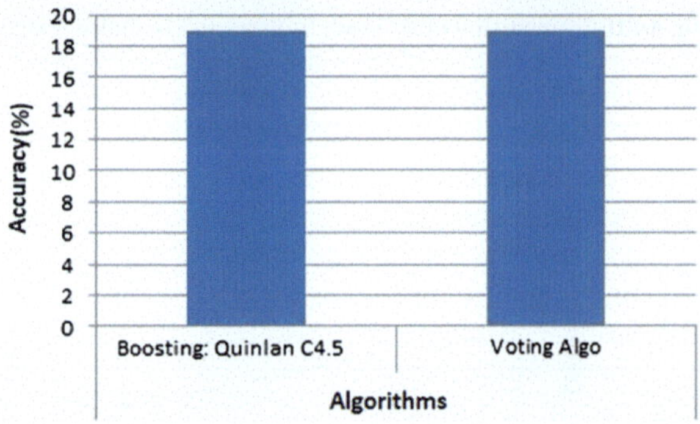

Fig. 6.13 Boosting versus voting algorithm with respect to accuracy

several versions into the voting algorithms, it was observed that the voting algorithm beats bagging and boosting in both stable and unstable procedures.

C. ***Comparative Analysis Between Basic and Ensemble Classifiers***

- ***SVM Versus Random Forest***

The work [50] explains how to identify EEG motor imagery signals using Filter Bank Common Spatial Patterns (FBCSP) and random forest. When a person imagines moving his hands, this EEG activity is produced. After selecting appropriate frequency bands, the FBCSP algorithm extracts usable characteristics (for imaging movement) and then classifies those valuable features using the classifier (random forest). In terms of accuracy and kappa value, we compare random forest and SVM in the graph below (used for accuracy measurement in multiple subjects).

Based on the accuracy measures shown in Figs. 6.14 and 6.15, we can state that random forest produces superior classification results than support vector machine. SVM has a higher error rate than random forest since it is susceptible to irrelevant feature selection (Fig. 6.15).

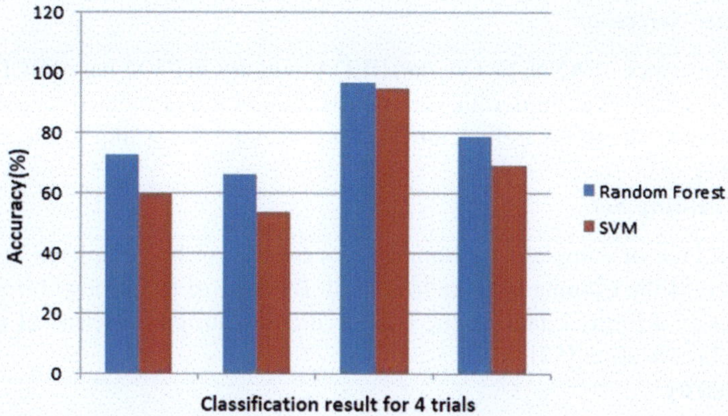

Fig. 6.14 Comparison between random forest and SVM with four tests

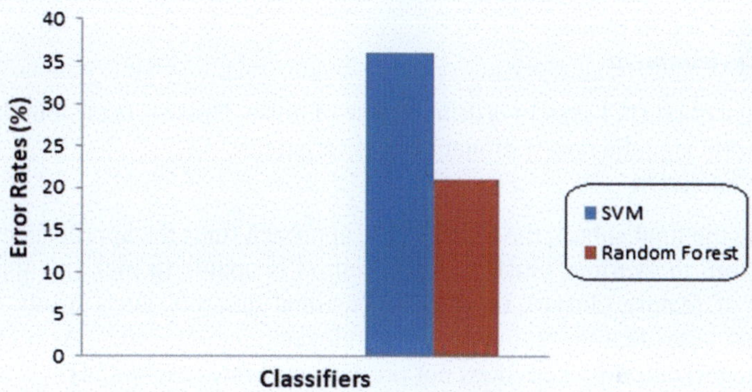

Fig. 6.15 Analysis of error rates between SVM and random forest

6.4 Open Research Problems

- ***Selecting Proper Brain Signal***

  Changes in brain signal over time cause various patterns, according to the author of paper [37]. Some patterns recognized by SSVEP signal could not be identified by ERD-based BCI. Paralyzed people can sometimes struggle to construct a correct pattern utilizing brain impulses. As a result, picking the right brain signal is crucial.
- ***Subject Reliance***

  FBCSP and random forest-based categorization do not produce the best results in time segments for each feature. The accuracy of classification is strongly reliant on the subject. The binary categorization of the EEG motor imagery signal is described in paper [50] that may fail to provide improved accuracy.

- ***Parameter Selection***

  The performance of ANN, FLDA, and BLDA methods in P300-based BCI systems is mostly reliant over numerous parameters such as experimented dataset, count of electrodes, electrode architecture, feature vectors, according to the author of paper [51].
- ***Small Training Set***

  Due to a lack of samples, the training process cannot be completed and tested effectively if the training set is rather small. Because of the strong correlation of attributes in a limited dataset, the genetic method suffers in terms of accuracy during classification [52].
- ***Nonlinearity***

  The brain is a nonlinear, highly complicated system [53]. Due to the nonlinear and dynamic character of the EEG signal, the nonlinear classifier has an advantage over the linear one [13]. The dynamic behavior of the signal can influence the feature extraction process.
- ***Noisy Environment***

  Building a good BCI system is difficult due of noise. Noise is occasionally caused by the brain signal's non-stationary characteristic.
- ***Feature Selection***

  To create the final subset, the FCTFS [54] approach finds the appropriate features subset from the various clusters. This method is unable to pick the appropriate number of feature clusters in high-dimensional datasets. As a result, the final subset excludes feature redundancy and relevant characteristics. As a result, the final categorization process does not produce accurate results [54].

6.5 EEG Signal-Based Emotional Data Classification

Emotion detection is a major research area in "brain–computer interfacing", where scientists are attempting to identify people's varied emotional states. The EEG signal is commonly utilized to detect many types of emotions. Multiple electrode channels capture the EEG signal, but only a handful of them are helpful for emotion identification. For dimension reduction, we use an approach called "correlation-based subset selection" in our research. The categorization process is then carried out utilizing the "Statistics with Higher Order" attributes of the minimized set of channels. In our research, four types of emotional states are playing roles, good, negative, furious, and harmonious. The proposed approach has an execution time of $O(n^2 + 2n)$. With a smaller number of channels, this model's classification accuracy is 82%.

6.5.1 Introduction

Emotion has a significant part in human–human interactions in real life. Given the changing nature of machines, we anticipate human–machine contact via BCI in the near future. The brain–computer interface (BCI) is a method of communicating between the brain and the outside environment without the use of physical movement. P300 speller, controlling wheel chair movement, motions of robotic arm, positioning of cursor, and more BCI applications are available. This study aims on a special correlation-based reduction technique of dimension for input channels and classifies them into four various categories of emotional states (shown in Fig. 6.15). According to this Fig. 6.16, we have chosen four major emotional states for the purpose of simplicity [48]. The positive, negative, furious, and harmony emotions are chosen from the arousal–valence model. However, emotion-specific channels for each class are chosen after conducting the dimension reduction method, which results in a true prediction rate on the validation data and good classification accuracy.

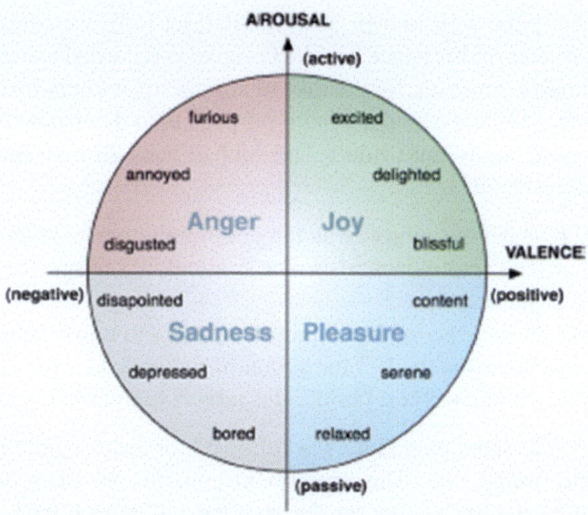

Fig. 6.16 Two-dimensional emotional model

Table 6.3 Summary of experimental arrangement

# of subjects	Sex (M)	Sex (F)	Total input channels	Total necessary channels	Sampling frequency (Hz)
32	18	18	Thirteen	Four	512

6.5.2 *Methodology*

6.5.2.1 Experimental Protocol

The investigation involves 32 healthy people ranging in age from 19 to 37 years old. Participants have been notified of the experiment's previous directives before the real experiment begins (trial at every step). The data is primarily collected using thirteen channels of electrodes of the cortex (FP1, FP2, F3, F4, Fz, PO3, P3, P4, Pz, O1, C3, T7, T8) that are used to store the raw signal during the experiment. The employed EEG signal has a sampling rate of 512 Hz. Table 6.3 has a detailed description of the experimental technique.

6.5.2.2 Experimental Setup

Using the well-known "DEAP" dataset [55], we categorized distinct moods of the subjects in this paper. The EEG signal is recorded using 32 electrodes in the "DEAP" dataset; however, for the sake of simplicity, we only investigate 13 electrode channels here. The experiment's main purpose is to determine human emotional states using visual emotional stimuli. The subject goes through the following processes in each experiment,

(1) Have the subject watch a one-minute music video.
(2) The respondent takes time is about 40 s to rate the video with respect to arousal and valence.
(3) A one-second black picture appears to rejuvenate the brain.
(4) Repeat steps 1–3 for a total of five videos.
(5) After a 4-s rest period, the subject is released from the experiment.

The same procedure was followed for the remaining subjects. Figure 6.17 depicts the timing rate. The experiment contains 40 trials, with one movie used for the test set and the rest for the training set in each trial. For determining the total of training and validation sets, leave-one outcross-validation (LOOCV) approach is used. Using arousal, valence, and dominance ratings, we were able to capture the various emotional responses of subjects during this experiment. The self-assessment manikin (SAM) [56] provides the arousal, valence, and dominance ratings. Table 6.4 provides a detailed description of the experimental setting. In Fig. 6.18, we have plotted the SAM.

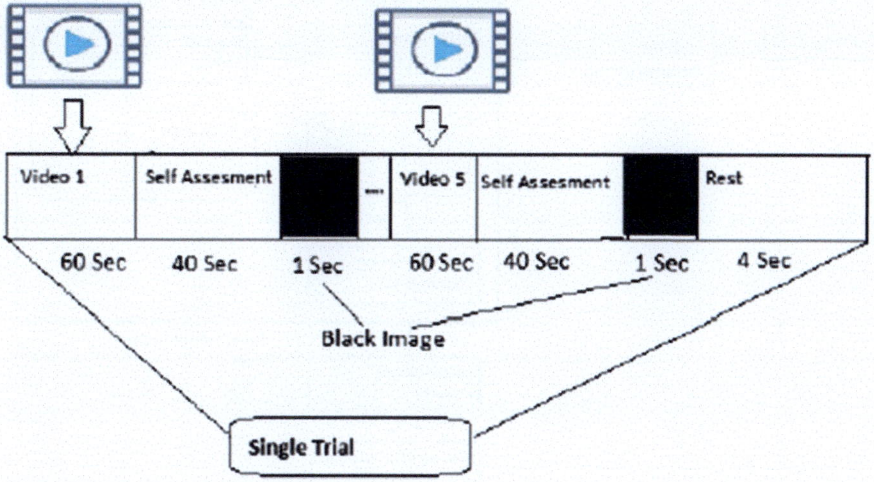

Fig. 6.17 Timing diagram per trial

Table 6.4 Necessary setup for the experiment

Trials #	Videos per trial	# of training set trails	# of validation set trails	Time taken per trial (s)
Forty	Five	165	45	510

Fig. 6.18 Manikin with self-assessment from valence SAM, arousal SAM, dominance SAM

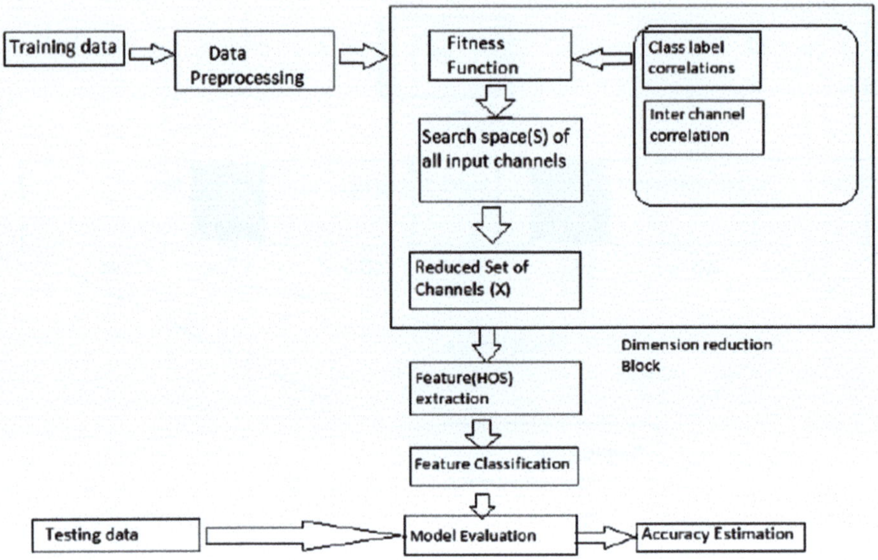

Fig. 6.19 Flowchart of the study

6.5.2.3 Framework of the Study

Not all the channels in our brain cortex are suitable for detecting emotions. As a result, we suggested a new correlation-based channel dimension minimization technique. By using the smaller set of channels, we extract the key properties linked to emotion classification. However, a linear discriminant analysis (LDA) classifier is employed in this research as the output class variance is similar and variables are categorical types. Figure 6.19 depicts a flowchart of the complete planned project.

6.5.2.4 Emotional States Classification with Correlation-Based Dimension Minimization

We have collected the brain activity using electrodes of 13 channels and put it inside a database in our proposed work. Only a few of the 13 channels are suitable for classification of emotional states. We have stored the amplitude (μv) values of the entire necessary channels for a specified interval of time in the database. The filtered data is pushed into our introduced reduction technique for fetching the necessary channels after the data preprocessing activity.

For each channel, we evaluate the objective functions and include the channel with the largest fitness value. The method of picking the best fitness value-based channel and inserting into the minimized subset is running until the subset has all of the required channels and no changes are made. Once the minimized subset is generated, we can obtain HOS features from it and move on to the classification of emotions process. The "Pearson correlation" coefficient is used to calculate the objective function (F_c) (Eq. 6.1).

$$\rho_{v1,v2} = \frac{\mathrm{cov}(v1, v2)}{\sigma_{v1}\sigma_{v2}} \tag{6.1}$$

whereas cov($v1$, $v2$) is the covariance in pair of channels, and $\sigma v1$ and $\sigma v2$ are standard deviation of $v1$ and $v2$ channels, respectively.

$$F_c = \frac{n.C_{fc}}{\sqrt{\rho_{v1,v2}}} \tag{6.2}$$

where n is the total number of channels, C_{fc} is the connection between feature classes, and $\rho_{v1,v2}$ is the correlation in pair of channels.

The proposed input channel dimension reduction technique is presented in Algorithm 1.

Similarity matrix: represent inter-channel correlation

$$\mathrm{Mat}_{\mathrm{Sim}} = \begin{array}{c} \\ P3 \\ FP2 \\ F3 \\ F4 \\ T7 \\ T8 \\ C3 \\ F_z \\ FP1 \\ P_z \\ PO3 \\ O1 \\ P4 \end{array} \begin{array}{c} \begin{array}{ccccccccccccc} P3 & FP2 & F3 & F4 & T7 & T8 & C3 & F_z & FP1 & P_z & PO3 & O1 & P4 \end{array} \\ \left[\begin{array}{ccccccccccccc} 1 & 0.77 & 0.90 & 0.11 & 0.20 & 0.01 & 0.26 & 0.89 & 0.80 & 0.95 & 0.36 & 0.09 & 0.94 \\ & 1 & 0.9 & 0.14 & 0.16 & 0.019 & 0.30 & 0.86 & 0.89 & 0.70 & 0.51 & 0.21 & 0.85 \\ & & 1 & 0.93 & 0.40 & 0.21 & 0.45 & 0.95 & 0.91 & 0.78 & 0.27 & 0.049 & 0.90 \\ & & & 1 & 0.79 & 0.81 & 0.87 & 0.93 & 0.40 & 0.32 & 0.66 & 0.38 & 0.29 \\ & & & & 1 & 0.89 & 0.81 & 0.46 & 0.41 & 0.43 & 0.38 & 0.63 & 0.42 \\ & & & & & 1 & 0.77 & 0.27 & 0.23 & 0.35 & 0.46 & 0.57 & 0.24 \\ & & & & & & 1 & 0.44 & 0.53 & 0.43 & 0.64 & 0.33 & 0.45 \\ & & & & & & & 1 & 0.85 & 0.82 & 0.86 & 0.23 & 0.92 \\ & & & & & & & & 1 & 0.80 & 0.36 & 0.008 & 0.93 \\ & & & & & & & & & 1 & 0.084 & 0.20 & 0.87 \\ & & & & & & & & & & 1 & 0.84 & 0.18 \\ & & & & & & & & & & & 1 & 0.15 \\ & & & & & & & & & & & & 1 \end{array}\right] \end{array}$$

(correlation (FP1, FP2) = correlation (FP2, FP1)). The similarity matrix is symmetric. As a result, the similarity matrix has the shape of an upper triangular matrix. We compute the fitness function and produce the subset of reduced channels after forming the similarity matrix of inter-channel correlation. Figure 6.20 depicts the development of the reduced subset (X). The most optimal channel in each run was highlighted in green, and after the fourth run, we verified that all of the needed emotion detection channels (FP1: negative, F3: angry, T7: positive, and PO3 for harmony emotion) had been included to the subset (X), therefore finally terminated the procedure [57].

In Algorithm 1, we have discussed our proposed dimension minimization process. At the beginning stage, all the input channels (n) are considered and selected optimum number of channels through the objective function. In each iteration, root channel

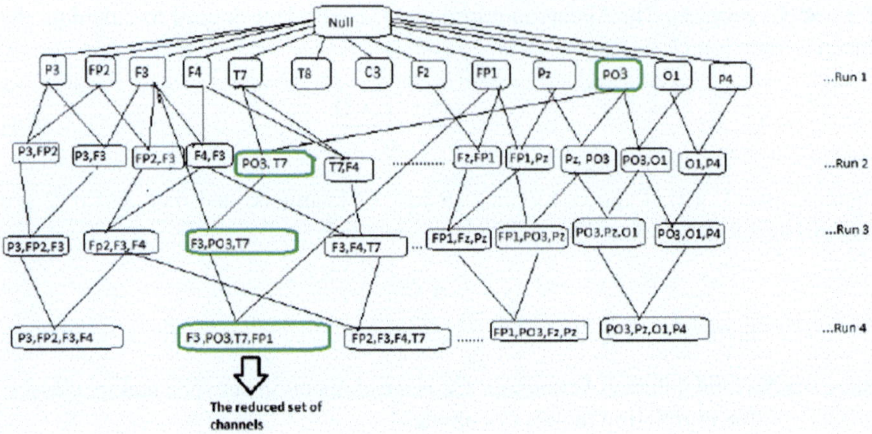

Fig. 6.20 Result of dimension reduction technique

(r) has been considered as the most effective channel. When X = NULL in the first run, the fitness function is solely determined by the input channel and feature class correlation. Finally, the process is ended, and the smallest set of channels (X) is returned. After our suggested algorithm (CSS) reduces the dimension, we fetch the necessary higher-order statistics characteristics from the smaller set of channels and proceed to emotion classification. Figure 6.21 shows the actual locations of placing electrodes [58, 59].

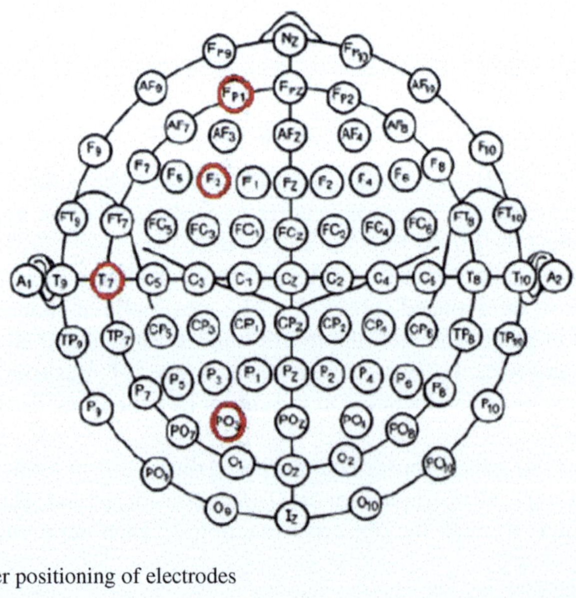

Fig. 6.21 Proper positioning of electrodes

Algorithm 1: Dimension reduction of input channels using CSS.

Input : Number of input channels(n), amplitude(μv) of input channels.

Output: Reduced set(X) of channels.

Search space(S): Contains all the input channels(n) with amplitude value for a specific time instance.
|S|=n and |X|=NULL.
Fitness function(F_c)=$\frac{n.C_{fc}}{\sqrt{\rho_{v1,v2}}}$
root_channel(r) = NULL.
while *((channels in the X has changed) AND (all emotion-specific channels are not present in X))* **do**
  foreach *channel i in S* **do**
    Calculate feature class correlation/entropy function(C_{fc}) of i^{th}channel.
    if *(r == NULL)* **then**
      $F_c(i)$ = n.$C_{fc}(i)$
    else
      Compute the inter-channel correlation between r and other (n-1) channels.
      Calculate the Fitness function $F_c(i, M)$, where M $\in$ (i+1,i+2,...n-2,n-1)
    end
  end
  Choose the channel(i) having Maximum Fitness value.
  r $\leftarrow$ i.
  X= X $\cup$ {i}.
end
return X;

6.5.3 *Result Analysis*

The following computing platform is used for the implementation purpose.

SignalPlant (version 3.3.2), R platform [60].

Hardware Premises: Windows 7 (64 bit) operating system, Intel Core (TM) i3 processor, 4 GB RAM, 2.26 GHz clock speed.

6.5.3.1 Preprocessing

We utilized "SignalPlant" software [60] to preprocess the raw EEG data (deletion of artifacts, filtering, etc.). The raw EEG signal was subjected to "band-pass filtering" in order to remove noise. Figure 6.22 shows the raw and filtered EEG signals. After the EEG signal has been filtered, we can go on to the input channel dimension reduction

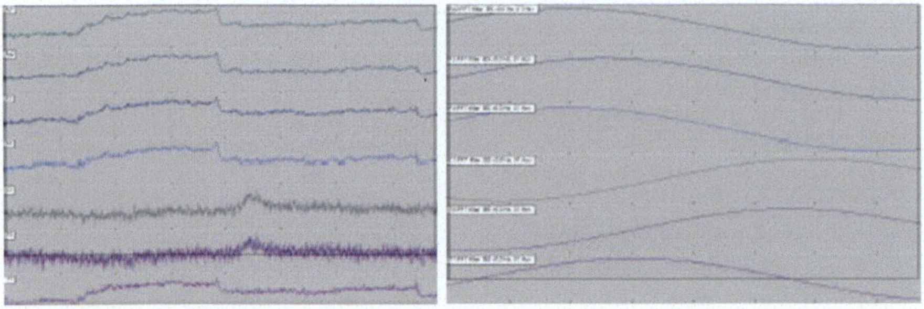

(a) (b)

Fig. 6.22 **a** Waveform of raw EEG signal. **b** Modified EEG signal form after filtering

approach. Then, using the decreased number of channels, we fetch the useful features of HOS and do classification.

6.5.3.2 Analysis of Feature Minimization Technique

The fitness function is used to reduce the number of input channels in our suggested technique (correlation-based subset selection). We can see from Fig. 6.23 that utilizing our suggested approach and the smallest number of channels, we get the best classification accuracy. Because our suggested system minimized duplicate channels, so we receive greater accuracy. The anticipated time of our suggested approach is plotted versus the number of input channels in Fig. 6.24.

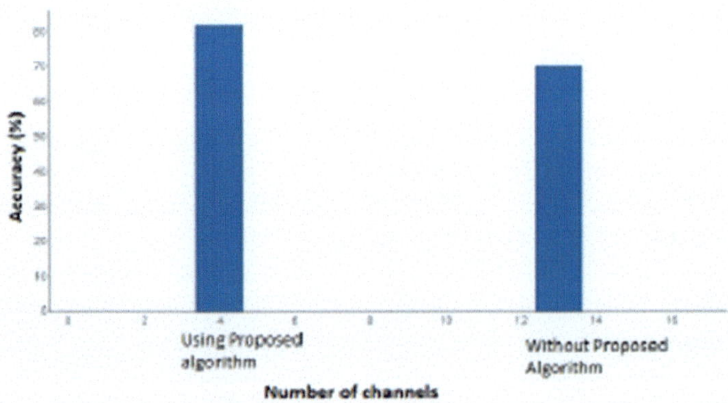

Fig. 6.23 Performance analysis of classification

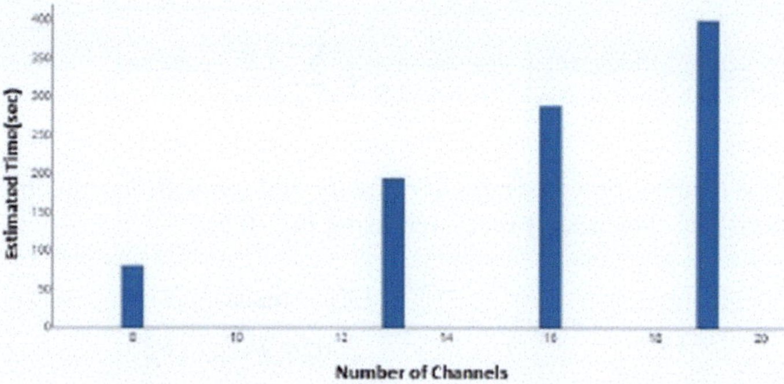

Fig. 6.24 Calculated time vs. number of channels

6.5.3.3 Time Complexity Analysis

- ***Complexity Analysis of Time***

  Inter-channel correlations: $O(n^2)$,

  Attribute class correlation calculation: $O(n)$,

  Required time to choose channel with maximum objective function: $O(n)$,

  Overall complexity: $O(n^2 + 2n)$, where n = total number of channels.
- ***Complexity Analysis of Space***

  Total space complexity: $O(n)$, where "n" is the total number of channels.

  From the above analysis, we can conclude our proposed algorithm's time complexity in quadratic time. The space complexity is linear because each run can store the maximum "n" number of channels in memory.

6.5.3.4 Feature Extraction and Selection

- **Features Statistics**

  The mean is used to calculate the average amplitude value of the acquired EEG data samples within the designated area. It is evaluated in Eq. (6.3).

$$\text{mean} = \frac{1}{n_e - n_s} \sum_{i=n_s}^{n_e - n_s} X_i \tag{6.3}$$

where n_s, n_e is the initial point and last point of the EEG data. The total number of samples is calculated as $n_e - n_{s,}$ and X_i represents the vertical axis value of the curve. It is expressed in Eq. (6.4).

$$\text{stddev} = \sqrt{\frac{1}{(n_e - n_s) - 1} \sum_{i=n_s}^{n_e - n_s} (X_i - \text{mean})^2} \tag{6.4}$$

Skewness is a metric that measures asymmetry. It depicts the asymmetry of a data distribution. The skewness is measured using Eq. (6.5).

$$\text{Skewness} = \frac{\sum_{i=n_s}^{n_e} (X_i - \text{mean})^3 / N}{(\text{stddev})^3} \tag{6.5}$$

The total number of sample points is "N". When compared to the normal distribution, kurtosis determines whether the data is highly or weakly tailed. The existence of noise in the data is indicated by a high kurtosis value, while the absence of outliers is shown by a low kurtosis value. The kurtosis is measured using Eq. (6.6).

$$\text{Kurtosis} = \frac{\sum_{i=n_s}^{n_e} (X_i - \text{mean})^4 / N}{(\text{stddev})^4} \tag{6.6}$$

We execute categorization of usable channels once the important statistical features have been determined (FP1, F3, T7, PO3). We use a leave-one-out cross-validation approach to evaluate our model [61].

LDA Classification and Its Analysis

We use a confusion matrix to examine the performance of the classification process in this paper. The confusion matrix is a table that shows how well a classification model performs on a set of test data for which the true values are already known. The confusion matrix summarizes the number of correct and wrong predictions for each class. We can easily see how much our classification model becomes "confused" during predictions with the help of the confusion matrix.

Confusion Matrix:

$\text{Mat}_{\text{con}} =$

	Positive Emotion	Negative Emotion	Angry	Harmony
Positive Emotion	32	42	4	4
Negative Emotion	14	611	3	5
Angry	17	31	18	10
Harmony	18	30	7	163

Table 6.5 and Fig. 6.25 show that emotional state "harmony" has the best sensitivity, whereas emotional state "positive feeling" (class 1) has the highest specificity. Class 2 denotes the highest level of accuracy using the largest $F1$-measure. On the basis of three accuracy parameters, we have shown the statistical performance of multiple class classification in Fig. 6.25. According to Fig. 6.26, our model provides 82% of average acceptable accuracy for the balanced dataset.

Table 6.5 Statistical performance analysis

Statistical measures	"+ve" emotion	"−ve" emotion	Anger state	Harmony state
Sensitivity	38.95	86.02	55.99	90.08
Precision	38.89	97.03	24.17	75.10
Specificity	95	93.00	93.91	92.95
Prevalence	8.13	71.11	3.27	17.91
$F1$-measure	27.99	42.98	5.29	4.96

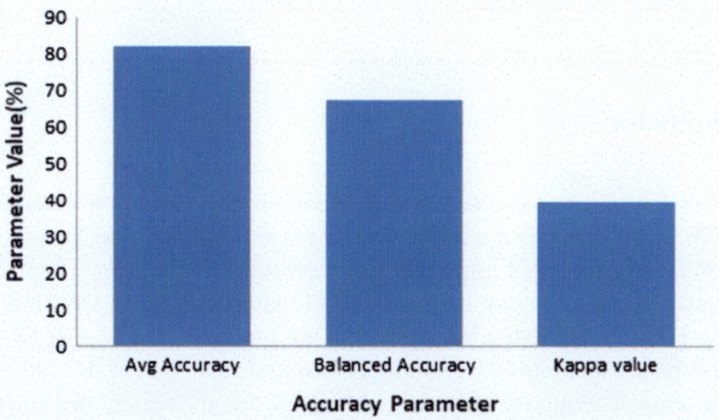

Fig. 6.25 Classifiers performance analysis based on accuracy parameters

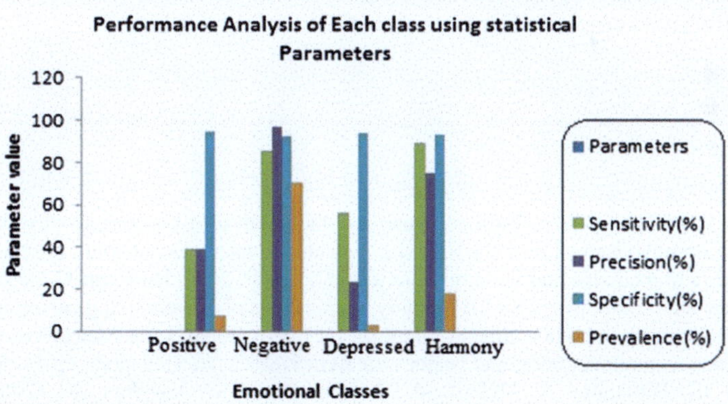

Fig. 6.26 Statistical performance analysis

6.5.3.5 Discussion

For categorization, we used higher-order statistical variables (mean, skewness, kurtosis, and so forth). The processing of signal and statistical analysis are carried out using the "SignalPlant" software [60]. The confusion matrix was used to analyze various classification characteristics (sensitivity, specificity, accuracy, and prevalence). We evaluated the performance of our anticipated model using those parameters. Our model's classification accuracy is 82%, which is significantly higher than that of other emotional detection models [6, 30, 53]. The classification process' accuracy is determined by a variety of criteria, including the type of signals, the subject numbers, the subjects' attentiveness, the location of electrodes, and so on [62]. Though we have used emotion categorization on able-bodied subjects, we believe that in the near future, we will be able to use it on disabled persons as well. As a result, paralyzed persons will be able to convey their feelings over a certain period of time, increasing the usefulness of the BCI system.

6.6 Conclusion

We studied distinct human emotional states using higher-order statistical factors in this study. We have developed a novel dimension minimization technique for input channels to classify the most important emotion-related channels. Therefore, from those relevant channels, we extract useful HOS features and do LDA classification. Our proposed dimension reduction algorithm has an execution time of $O(n^2 + 2n)$. Using the proposed technique, we did a performance analysis of the categorization process (dimension reduction process). After several measures, the classification accuracy is calculated of the proposed model and finally show the comparison with other state-of-the-art models. However, we plan to use different SVM kernels to construct a classification procedure for a high volume of nonlinear data and determine the most effective kernel for emotion classification in the future.

References

1. Darvishi S, Al-Ani A (2007) Brain-computer interface analysis using continuous wavelet transform and adaptive neuro-fuzzy classifier. In: 29th annual international conference of the IEEE engineering in medicine and biology society, 2007. EMBS 2007. IEEE, pp 3220–3223
2. Chakladar DD, Chakraborty S (2017) Study and analysis of a fast moving cursor control in a multithreaded way in brain computer interface. In: International conference on computational intelligence, communications, and business analytics. Springer, Singapore, pp 44–56
3. Wolpaw JR, Birbaumer N, McFarland DJ, Pfurtscheller G, Vaughan TM (2002) Brain–computer interfaces for communication and control. Clin Neurophysiol 113(6):767–791
4. Schwenk H, Bengio Y (2000) Boosting neural networks. Neural Comput 12(8):1869–1887
5. Roman-Gonzalez A (2012) EEG signal processing for BCI applications. In: Human–computer systems interaction: backgrounds and applications, vol 2. Springer, Berlin, pp 571–591
6. Bashar SK, Hassan AR, Bhuiyan MIH (2015) Identification of motor imagery movements from EEG signals using dual tree complex wavelet transform. In: 2015 international conference on advances in computing, communications and informatics (ICACCI). IEEE, pp 290–296
7. Yoon HJ, Chung SY (2013) EEG-based emotion estimation using Bayesian weighted-log-posterior function and perceptron convergence algorithm. Comput Biol Med 43(12):2230–2237

8. Rejer I (2015) Genetic algorithm with aggressive mutation for feature selection in BCI feature space. Pattern Anal Appl 18(3):485–492
9. Wang XW, Nie D, Lu BL (2014) Emotional state classification from EEG data using machine learning approach. Neurocomputing 129:94–106
10. McFarland DJ, Anderson CW, Muller KR, Schlogl A, Krusienski DJ (2006) BCI meeting 2005-workshop on BCI signal processing: feature extraction and translation. IEEE Trans Neural Syst Rehabil Eng 14(2):135–138
11. Fatehi TAH, Suleiman ABR (2011) Features extraction techniques of EEG signals for BCI application
12. Vidaurre C, Kramer N, Blankertz B, Schlogl A (2009) Time domain parameters as a feature for EEG-based brain-computer interfaces. Neural Netw 22(9):1313–1319
13. Li Y, Ma Z, Lu W, Li Y (2006) Automatic removal of the eye blink artifact from EEG using an ICA-based template matching approach. Physiol Meas 27(4):425
14. McFarland DJ, Krusienski DJ, Wolpaw JR (2006) Brain-computer interface signal processing at the Wadsworth Center: mu and sensorimotor beta rhythms. Prog Brain Res 159:411–419
15. Gao JF, Yang Y, Lin P, Wang P, Zheng CX (2010) Automatic removal of eye-movement and blink artifacts from EEG signals. Brain Topogr 23(1):105–114
16. Chen LL, Madhavan R, Rapoport BI, Anderson WS (2013) Realtime brain oscillation detection and phase-locked stimulation using autoregressive spectral estimation and time-series forward prediction. IEEE Trans Biomed Eng 60(3):753–762
17. Li Y, Long J, Yu T, Yu Z, Wang C, Zhang H, Guan C (2010) An EEG-based BCI system for 2-D cursor control by combining Mu/Beta rhythm and P300 potential. IEEE Trans Biomed Eng 57(10):2495–2505
18. Muller KR, Anderson CW, Birch GE (2003) Linear and nonlinear methods for brain-computer interfaces. IEEE Trans Neural Syst Rehabil Eng 11(2):165–169
19. Chiappa S, Barber D (2006) EEG classification using generative independent component analysis. Neurocomputing 69(7):769–777
20. Lemm S, Blankertz B, Dickhaus T, Müller KR (2011) Introduction to machine learning for brain imaging. Neuroimage 56(2):387–399
21. Huan NJ, Palaniappan R (2004) Neural network classification of autoregressive features from electroencephalogram signals for brain–computer interface design. J Neural Eng 1(3):142
22. Hsu WY, Lin CC, Ju MS, Sun YN (2007) Wavelet-based fractal features with active segment selection: application to single-trial EEG data. J Neurosci Methods 163(1):145–160
23. Kolodziej M, Majkowski A, Rak R (2011) A new method of EEG classification for BCI with feature extraction based on higher order statistics of wavelet components and selection with genetic algorithms. Adaptive and Natural Computing Algorithms, 280–289. Lakshmi MR, Prasad DT, Prakash DVC (2014) Survey on EEG signal processing methods. Int J Adv Res Comput Sci Softw Eng 4(1)
24. Hsu WY (2010) EEG-based motor imagery classification using neuro-fuzzy prediction and wavelet fractal features. J Neurosci Methods 189(2):295–302
25. Lawhern V, Hairston WD, McDowell K, Westerfield M, Robbins K (2012) Detection and classification of subject-generated artifacts in EEG signals using autoregressive models. J Neurosci Methods 208(2):181–189
26. McFarland DJ, Krusienski DJ, Sarnacki WA, Wolpaw JR (2008) Emulation of computer mouse control with a noninvasive brain–computer interface. J Neural Eng 5(2):101
27. Boye AT, Kristiansen UQ, Billinger M, do Nascimento OF, Farina D (2008) Identification of movement-related cortical potentials with optimized spatial filtering and principal component analysis. Biomed Sig Process Control 3(4):300–304
28. Rakotomamonjy A, Guigue V (2008) BCI competition III: dataset II ensemble of SVMs for BCI P300 speller. IEEE Trans Biomed Eng 55(3):1147–1154
29. Garrett D, Peterson DA, Anderson CW, Thaut MH (2003) Comparison of linear, nonlinear, and feature selection methods for EEG signal classification. IEEE Trans Neural Syst Rehabil Eng 11(2):141–144

30. Blankertz B, Lemm S, Treder M, Haufe S, Muller KR (2011) Singletrial analysis and classification of ERP components—a tutorial. Neuroimage 56(2):814–825
31. Ramadan RA, Refat S, Elshahed MA, Ali RA (2015) Basics of brain computer interface. In: Brain-computer interfaces. Springer, Cham, pp 31–50
32. Murugappan M, Ramachandran N, Sazali Y (2010) Classification of human emotion from EEG using discrete wavelet transform. J Biomed Sci Eng 3(04):390
33. Garcia GN, Ebrahimi T, Vesin JM (2003) Support vector EEG classification in the Fourier and time-frequency correlation domains. In: First international IEEE EMBS conference on neural engineering, 2003. Conference proceedings. IEEE, pp 591–594
34. Ishfaque A, Awan AJ, Rashid N, Iqbal J (2013) Evaluation of ANN, LDA and decision trees for EEG based brain computer interface. In: 2013 IEEE 9th international conference on Emerging Technologies (ICET). IEEE, pp 1–6
35. Kayikcioglu T, Aydemir O (2010) A polynomial fitting and *k*-NN based approach for improving classification of motor imagery BCI data. Pattern Recogn Lett 31(11):1207–1215
36. Thomas E, Fruitet J, Clerc M (2013) Combining ERD and ERS features to create a system-paced BCI. J Neurosci Methods 216(2):96–103
37. Bhattacharyya S, Khasnobish A, Chatterjee S, Konar A, Tibarewala DN (2010) Performance analysis of LDA, QDA and *K*NN algorithms in left-right limb movement classification from EEG data. In: 2010 international conference on systems in medicine and biology (ICSMB). IEEE, pp 126–131
38. http://openvibe.inria.fr/datasets
39. Ting W, Guo-zheng Y, Bang-hua Y, Hong S (2008) EEG feature extraction based on wavelet packet decomposition for brain computer interface. Measurement 41(6):618–625
40. https://signalplant.codeplex.com/
41. James G, Witten D, Hastie T, Tibshirani R (2013) An introduction to statistical learning, vol 112. Springer, New York
42. Hoffmann U, Garcia G, Vesin JM, Diserens K, Ebrahimi T (2005) A boosting approach to P300 detection with application to brain-computer interfaces. In: 2nd international IEEE EMBS conference on neural engineering, 2005. Conference proceedings. IEEE, pp 97–100
43. Seeck M, Koessler L, Bast T, Leijten F, Michel C, Baumgartner C, Beniczky S (2017) The standardized EEG electrode array of the IFCN. Clin Neurophysiol 128(10):2070–2077
44. Akram F, Han SM, Kim TS (2015) An efficient word typing P300-BCI system using a modified T9 interface and random forest classifier. Comput Biol Med 56:30–36
45. Dietterich TG (2000) An experimental comparison of three methods for constructing ensembles of decision trees: bagging, boosting, and randomization. Mach Learn 40(2):139–157
46. Xing-Yu W, Jing J, Zhang Y, Bei W (2013) Brain control: human computer integration control based on brain-computer interface approach. Acta Automat Sin 39(3):208–221
47. Suleiman ABR, Fatehi TAH (2007) Features extraction techniqes of EEG signal for BCI applications. Faculty of Computer and Information Engineering, Department College of Electronics Engineering, University of Mosul, Iraq
48. Bauer E, Kohavi R (1999) An empirical comparison of voting classification algorithms: bagging, boosting, and variants. Mach Learn 36(1):105–139
49. Breiman L (1996) Bagging predictors. Mach Learn 24(2):123–140
50. Bentlemsan M, Zemouri ET, Bouchaffra D, Yahya-Zoubir B, Ferroudji K (2014) Random forest and filter bank common spatial patterns for EEG-based motor imagery classification. In: 2014 5th international conference on intelligent systems, modelling and simulation (ISMS). IEEE, pp 235–238
51. Bakhshi A, Ahmadifard A (2012) A comparison among classification accuracy of neural network, FLDA and BLDA in P 300-based BCI system. Int J Comput Appl 46(19):11–15
52. Duda RO, Hart PE, Stork DG (2001) Pattern recognition, 2nd edn. Wiley Interscience
53. Abdulkader SN, Atia A, Mostafa MSM (2015) Brain computer interfacing: applications and challenges. Egypt Inf J 16(2):213–230
54. Goswami S, Das AK, Chakrabarti A, Chakraborty B (2017) A feature cluster taxonomy based feature selection technique. Expert Syst Appl 79:76–89

55. http://www.eecs.qmul.ac.uk/mmv/datasets/deap/index.html
56. Bradley MM, Lang PJ (1994) Measuring emotion: the self-assessment manikin and the semantic differential. J Behav Ther Exp Psychiatry 25(1):49–59
57. Dey N (ed) (2016) Classification and clustering in biomedical signal processing. IGI Global
58. Selvaraj J, Murugappan M, Wan K, Yaacob S (2013) Classification of emotional states from electrocardiogram signals: a non-linear approach based on hurst. Biomed Eng Online 12(1):44
59. Shi F, Dey N, Ashour AS, Sifaki-Pistolla D, Sherratt RS (2019) Meta-KANSEI modeling with valence-arousal fMRI dataset of brain. Cogn Comput 11(2):227–240
60. Sitaram R, Zhang H, Guan C, Thulasidas M, Hoshi Y, Ishikawa A, Birbaumer N (2007) Temporal classification of multichannel near-infrared spectroscopy signals of motor imagery for developing a brain-computer interface. Neuroimage 34(4):1416–1427
61. Liu J, Dey N, Das N, Crespo RG, Shi F, Liu C (2022) Brain fMRI segmentation under emotion stimuli incorporating attention-based deep convolutional neural networks. Appl Soft Comput 122:108837
62. Shi X, Xu G, Shen F, Zhao J (2015) Solving the data imbalance problem of P300 detection via random under-sampling bagging SVMs. In: 2015 international joint conference on neural networks (IJCNN). IEEE, pp 1–5

Part III
Quantum Computing for Data Analysis

Chapter 7
Quantum Computing in Machine Learning

7.1 Introduction

For physicists, statisticians, computer engineers, and other technical professionals, quantum information processing is an essential technology with many outstanding challenges [1]. This chapter gives a brief overview of some well-known previous works on the areas of quantum AI, quantum ML, and quantum image handling. The primary goal of this work is to compile the current state of the art in various quantum image handling tasks, storage techniques along with various quantum supervised and unsupervised learning algorithms and summarize the recent developments in this field. Other from that, several of the available quantum approaches are examined based on their applications, distinctions, limitations, and potential uses. In this chapter, we study and describe an overview of how quantum principles can contribute to improving the performance of different machine learning and image processing techniques. This chapter also discusses some significant applications (such as quantum image compression [2], quantum image edge detection [3], quantum Histogram, quantum ghost imaging [4]) and also combines with a set of popular natural computing techniques (like artificial intelligence, artificial neural networks [5], ant colony optimization, etc.). However, it also provides the advantages of using those methods in quantum systems compared to classical systems. Finally, it discusses some open research issues and challenges in this area that will boost up the future researchers of this field. Quantum computing is integrated with natural algorithms such as quantum genetic algorithm (quantum computing and genetic algorithm), quantum neural network, ant colony-based quantum computing, quantum clustering, quantum cellular automata, quantum classification, quantum backpropagation, quantum inspired evolutionary algorithm [6] to offer several advantages in the areas of data mining, machine learning, and optimization processes. The main aim of the quantum researcher is to discover more quantum algorithms on AI domain providing notable speedup over existing classical AI algorithms. Besides that, AI researchers also agree that quantum community has enough potential for finding the solutions to some currently complex (NP-hard) problems [7, 8]. Machine learning, which was discussed

S. Chakraborty and L. Dey, *Computing for Data Analysis: Theory and Practices*, Data-Intensive Research, https://doi.org/10.1007/978-981-19-8004-6_7

in the preceding section, is maybe the only arena where quantum computation and intelligent systems have already interacted in a productive way. Decision problems described by a family of decision trees can be addressed exponentially quicker with quantum computers and decision tree exploration than with classical random walks. When AI combines to quantum processing, it leads to play several roles such as quantum optimization search, quantum gaming, quantum intelligent learning. Quantum parallel processing, entanglement, teleportation, and other techniques can aid in the cheaper and more effective exploration of various aspects of AI. The brain, for example, can be seen as a system made up of multiple entangled subsystems. This entanglement may illustrate why, despite the absence of a single subsystem's output, it may still be capable of determining it by integrating information from other subsystems [7]. In semantic analysis and, more broadly, in AI, quantum mechanics plays a key role. Some semantic characteristics of natural languages, such as ambiguity, can be described in quantum theory using quantum superposition. Natural language processing is accelerated up as a result. Quantum entanglement is crucial in identifying essential word relationships in natural languages. More experimental study is needed, in particular, to identify the effectiveness of employing quantum theory in the field of AI. Quantum machine learning (QML) is a study of implementing machine learning algorithms in quantum systems with the help of quantum properties such as superposition, entanglement, parallelism to solve learning problems. We can consider it as a part of quantum AI. Quantum researchers have developed few techniques like quantum annealing, quantum walks, quantum phase estimation, quantum Fourier transform, amplitude amplification, etc. that exploit the characteristics of quantum computers over the field of machine learning. These help to explain the alternative algorithms to make decisions understandable by a human. QML is able to handle optimization that is not convex in the area of machine learning along with some convex optimization problems to some extent. One of the problems related to classical machine learning is to handle huge amounts of data. Quantum systems, on the other hand, have a Hilbert space whose dimension grows exponentially in proportion to the size of the system, allowing them to hold and modify massive amounts of data. However, an important concern is efficiently and precisely initializing this quantum state with the desired data. Quantum information processing also allows for parallel processing and quick search (grover's) procedures over the data from where essential information or patterns can be mined. Ezhov [9] and Ventura [10] represented each basis state in a superposition as a labeled object of training data. Grover's algorithm is applied on the initial basis states. The current state of the system is measured, and the classification result is obtained after a certain number of iterations. Quantum walks work on a straight curve, cyclic structure, or n-bit cubic structure that searching or navigating between two different positions with exponential speedup. A research work [11] introduced the use of m number of one-dimensional discrete-time quantum walks (DTQW) on a d-regular bipartite graph and lattice to discover the closest data points among neighbors with exponential speedup and localization phenomenon. A paper [12] presented a quantum pattern recognition algorithm, which was more abstract than a statistical classification. A quantum version of a decision tree classifier, which can also deal with multiclass situations, is proposed in paper [13]. One of the very interesting works is which addressed association, rule mining in quantum settings and also proposed quantum association rule mining algorithm for the core part that supposed to search frequent itemsets from the candidate itemsets along with their support. A series of quantum techniques help it to do this operation with less time complexity compared to the classical counterpart [9]. HHL-based quantum linear system solving algorithms [14] and amplitude amplification obtain the rapid speedup over classical algorithm under certain conditions while implementing quantum counterparts of support vector machine (SVM), principle component analysis (PCA), and linear discriminant analysis (LDA). For k-means clustering, the Lloyd's quantum approach is suggested, which employs a novel variant of the quantum adiabatic algorithm to categorize M vectors into k clusters in time $O(k \log kMN)$ [15].

7.2 Quantum Hybrid Data Clustering

In this chapter, our main aim is to provide a novel hybrid quantum clustering technique by taking the help of some significant quantum principles, such as superposition of qubits, Shor's algorithm, Grover's algorithm, discrete quantum walks, that can work on one-dimensional regular graph. In our proposed algorithm, initially data points are navigating and randomly walk m-dimensional free space. With the help of the responses of other particles, supervision algorithm can select the correct neighbors and evaluate the transition probabilities to them. By inducing the probability distribution of particles, the quantum walk displays the potential placements of the particles. The particle positions are then modified as a result of the superposition state being assessed. We use an additional coin operation (hadamard/grover/fourier) and a shift operation in the finite case, as well as a regular (degrees of all vertices are the same), bipartite, undirected, reversible graph to find the future position. The quantum walk is the quantum equivalent of a classical random walk in quantum computing. Quantum walk is a valuable tool for creating efficient quantum algorithms with various quantum tools like Grover's amplitude amplification [6], entanglement, and so on. The walker's state in a classical random walk is represented by a probability density function across positions, but the walker's state in a quantum walk is defined by a superposition of states. Based on a probability distribution, the classical 1D random walk navigates either to the left with prob(p) or right with prob($1 - p$) along with the data particles. The expected distance of a quantum walk from the origin is greater than the classical random walk's expected distances [8]. In a classical walk, only constructive interference exists, whereas in a quantum walk, both constructive and destructive interferences exist. As a result, quantum walk has a quadratic speed advantage ($\sqrt{N}$) over classical random walk. Aside from that, the quantum walk is distinguished from the classical one by its reversible and unitary evolution. The discrete-time quantum walk works in a discrete space of locations using unitary coin tosses and positional shifting operations. In a discrete scenario, however, expanding the space of the physical system introduces an extra coin state. A particle moves simultaneously in this paradigm, i.e., with the same probabilities in both the left and right directions. When unitary operations are conducted, the particles will travel in the right and left directions due to the additional degree of coin state [6]. In this chapter, a d-regular bipartite reversible undirected graph is subjected to m 1D quantum random walks [16–18]. A discrete quantum walk's total state (H_t) is made up of the walker basis positions (H_p) and the basis position of a coin (H_c) residing in an infinite Hilbert space.

$$H_t = H_p \otimes H_c \tag{7.1}$$

However, the basic superposition state is represented as,

$$|\psi\rangle_{\text{init}} = |\text{position}\rangle_{\text{init}} \otimes |\text{coin}\rangle_{\text{init}} \tag{7.2}$$

where $|\text{coin}\rangle = \alpha|0\rangle_c + \beta|1\rangle_c$ and $|\alpha|^2 + |\beta|^2 = 1$. The coin ($C$) operation splits the walker's direction into a superposition of coin states, while the conditional shift (S) operation determines the walker's next position depending on the coin state. Therefore, the evolution operator (U) can be represented as,

$$U = S(C \otimes I) \tag{7.3}$$

This $C \otimes I$ performing only on the coin space $|C\rangle$ in a non-trivial way.

The basis states $|v\rangle$ in the Hilbert space Hv cover the vertices $v \in V$, while the degree basis states $|d\rangle$ cover the auxiliary Hilbert space H_A. On $H_A \otimes H_v$, the coin operator C and the shift operator S are performed unitarily, resulting in $S|a, v\rangle = |a, u\rangle$, where u and v are neighbors segregated by a certain distance. If the graph follows a cyclic structure having n points, therefore the quantum walk's Hilbert space is $C^2 \otimes C^n$. If we use the H operation as a coin-flipping operator, then the S operation can be written as,

$$S|\text{Right}, k\rangle = |\text{Right}, (k+1)\,\text{mod}(n)\rangle \tag{7.4}$$

$$S|\text{Left}, k\rangle = |\text{Left}, (k-1)\,\text{mod}(n)\rangle \tag{7.5}$$

However, the probability distribution can be received as,

$$P(x,t) = \sum_{i=1}^{\text{dim}} |\langle e_i, x|\psi(t)\rangle|^2 = \sum_{i=1}^{\text{dim}} |\langle x| \otimes \langle e_i|\psi(t)\rangle|^2 \tag{7.6}$$

where $|e_i\rangle$, $i = 1, 2, 3, \ldots, \text{dim}$. The proposed quantum clustering is divided into two categories. The first part of the algorithm will use the least distance-based technique to create an undirected bipartite network from the input information. Our suggested quantum algorithm requires this undirected bipartite graph as a prerequisite.

7.2.1 Methodology: Pseudo-steps of Proposed Quantum Clustering Process

Initialization:
Assume that our approach is used on a test dataset that has two initial growing clusters (using minimum distance measure matrices). The initial positions of the data particles are represented by these two clusters. These clusters are laid out in the form of a regular undirected bipartite network, with links between data points. To begin, choose two quantum registers to represent two clusters of the input graph $G = (V, E)$, where n is the total number of particles.
Consider the starting state $\psi(t)$ for the quantum walk operations after applying superposition according to the below equation.

$$|\psi(t)\rangle = \sum_{x,c} \alpha_{x,c}(t)|x, c\rangle \tag{7.7}$$

It is also referred as an initial state of the discrete quantum walk on the input graph G at time t.
Therefore, a distance function is considered as $d(.,.)$, and a predefined threshold value is considered as T_θ.
Now, the initial distance matrix is evaluated as $d(X_0^i, X_0^j)_{i,j=1,2,3,\ldots,n}$ and initial degree vector as $X_0^j(\text{deg})_{j=1,2,3,\ldots,n}$.
Repeat:
Evaluate the current distance matrix as $d(X_t^i, X_t^j)_{i,j=1,2,3,\ldots,n}$.
For i → 1 to n
    Discover the current neighbor sets $N_c(i)$
    Compute the current degree vector as $X_t^j(\text{deg})_{j=1,2,3,\ldots,n}$.
    Calculate the transition probabilities $P_t(i, j),\ j \in N_c(i)$ as,

$$P_{t,j\in C_2}(i, j) = \begin{cases} \frac{B_{t,j\in C_2}(i,j)}{\sum_{j\in C_2} B_t(i,j)}, & \text{if} j \in N_c(i), d(X_t^i, X_t^j) > \mu \\ 0, & \text{otherwise} \end{cases} \tag{7.8}$$

where $X_t^i \in C_1$ and $X_t^j \in C_2$.

$$B_t(i, j)_{j\in C_2} = \frac{(X_t^j(\text{deg})/\sum_{j\in C_{CO_2}} X_t^j(\text{deg})) \times (X_0^j(\text{deg})/\sum_{j\in C_2} X_0^j(\text{deg}))}{(d(X_t^i, X_t^j)) \times (d(X_0^i, X_0^j))} \tag{7.9}$$

where $N_c(i)$ denotes the neighbor set of a data point X_t^i; μ denotes the smallest distance between two data points; $X_t^j(\text{deg})$ is the current degree; and $X_0^j(\text{deg})$ is the initial degrees of a data point X_i, respectively. However, utilizing the current and beginning degrees of the same particle to determine the transition probability of each particle, the intermediate term $B_t(i, j)$ is necessary.

We can implement the coin transformation C as,

$$C = \begin{bmatrix} \sqrt{\rho} & \sqrt{1-\rho}e^{i\theta} \\ \sqrt{1-\rho}e^{i\phi} & -\sqrt{\rho}e^{i(\theta+\phi)} \end{bmatrix}$$

, and $U = S.(C \otimes I)$ is applied to the basic state r times in each dimension according to equations below. After $r = 1$ time, the updated state is represented as,

$$|\psi_1\rangle = \rho|+\rangle \otimes \left|\frac{l_{C_1}}{r}\right\rangle + \sqrt{1-\rho}e^{i\theta}|-\rangle \otimes \left|\frac{-l_{C_2}}{r}\right\rangle \tag{7.10}$$

after $r = 2$ times,

$$\begin{aligned} |\psi_2\rangle = \rho|+\rangle \otimes |l_{C_1}\rangle + \sqrt{\rho(1-\rho)}|-\rangle \otimes |(l_{C_1} - l_{C_2})/r\rangle + (1-\rho)|+\rangle \\ \otimes|(l_{C_1} - l_{C_2})/r\rangle - \sqrt{\rho(1-\rho)}|-\rangle \otimes |-l_{C_2}\rangle \end{aligned} \tag{7.11}$$

Measure the state and update the corresponding positions of each component of $X_t(i)$ according to the equations derived below.
($X^i_{t+1}(j) = X^i_t(j) + l_{C_1}$, if at position ($l_{C_1}$) or $X^i_{t+1}(j) = X^i_t(j) + (l_{C_1} - l_{C_2})/r$, if at position $(l_{C_1} - l_{C_2})/r$ or $X^i_{t+1}(j) = X^i_t(j) - l_{C_2}$, if at position $(-l_{C_2})$).
Where $l_{C_1} = \rho \times (X_{t,j\in C_1} - X^i_t(j))$, $j \in 1, 2, \ldots, m$, and
$l_{C_2} = \rho \times (X_{t,j\in C_2} - X^i_t(j))$, $j \in 1, 2, \ldots, m$.

Finally, calculate the sum of navigating distances of each data particle $D_i = \sum_{j=1}^{m} l_j$.
End For
Until $\sum_{i=1}^{n} D_i < T_\theta$

Output: The particle's probability distribution, which is induced by quantum walks, will be used to find the updated positions of the particles. Because data points navigate around in space at random, they progressively clump with each other and build distinct clusters.
Figure 7.1 depicts the whole quantum block diagram of the above quantum hybrid clustering technique. According to Fig. 7.1, the quantum comparator circuit computes the greatest transition probability (Max($Pt(i, j)$)) after computing the transition probabilities of all neighbor clusters. Therefore, the $|C\rangle$ and $|S\rangle$ are working together along with discrete-time quantum walks. Quantum parallelism is used to apply the whole unitary operation ($U = |C\rangle \otimes |S\rangle$) numerous times. The positions of l_{C_1}, l_{C_2} are then measured with various probabilities, and the positions of those particles are updated correspondingly. As part of our proposed quantum oracle, the entire procedure is running. Finally, in dynamic datasets, the quantum comparator circuit is employed to evaluate the correct placements of freshly inserted data. Table 7.1 represents a brief comparisons between the proposed technique and some reknown techniques in terms of clustering accuracy. This experiment uses a number of well-known public data sources collected from UCI and Kaggle repositories. From our experimental analysis, it is proved that the performance of the proposed approach is superior than some well-known methods in terms of clustering accuracy.

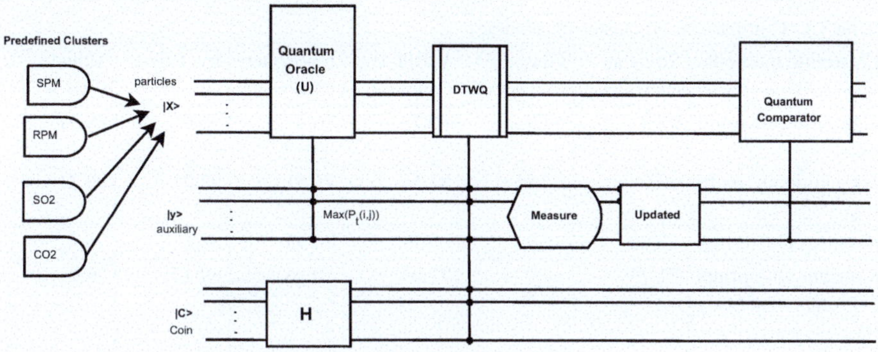

Fig. 7.1 Quantum circuit diagram of proposed approach

7.2.2 Analysis of Results

We chose four prominent datasets (soybean, glass, ionosphere, and iris) and a weather forecasting dataset, to examine the performance of the proposed approach on various real-world datasets (WBPCB collected from West Bengal Pollution Control Board). All of the data points, with their diverse characteristics, are dispersed throughout a high-dimensional space. The proposed quantum hybrid clustering technique is applied to the aforementioned datasets with varying k numbers of nearest neighbors, and the results are shown in Table 7.1. The suggested algorithm is written in MATLAB 13 and tested on Windows 10 with 8 GB of random access memory and intel i5 processor. The k-nearest neighbors of each data point in the dataset can be identified using that distance measure function. With respect to the soybean and iris datasets, 1D MCMS and LDA-K-means algorithms provide higher clustering accuracy than our suggested algorithm. All of the other scenarios, with the exception of those two, are win–win situations for our suggested method.

7.2.3 Computational Complexity Analysis

It takes $O(n)$ amount of computational time to build the undirected regular bipartite graph as principal clusters. After starting with any arbitrary data point from those two initial clusters, we must run for loops n times to reach each data point and finally compute their positions.

$$\sqrt{\sum_{i=1}^{n-2}(y_j - x_i)^2}, \quad j = 1, 2 \approx O(n). \tag{7.12}$$

Table 7.1 Comparison on clustering accuracy

Clustering methods	Soybean (%)	Glass (%)	WBPCB (%)	Ionosphere (%)	Iris (%)	Vehicle (%)
1D single-coin multi-step	92.3	62.9	80.7	76.1	95.0	88.91
1D multicoin multi-step	98.0	62.85	84.1	75.61	97.21	93.21
K-means clustering	67.89	46.97	69.79	70.93	88.8	76.3
K-means with principal component analysis	71.75	44.9	77.05	70.86	89.17	79.2
K-means with linear discriminant analysis	75.91	49.93	80.28	70.93	97.86	86.69
Proposed algorithm	94.9	64.79	86.0	77.8	93.59	94.39

Table 7.2 Comparisons with some state-of-the-art techniques

Smallest distance	AGNES	Density-based	K-means	Proposed clustering
$O(n)$	$O(n^2 \log(n))$, $O(2^n)$	$O(n^2)$	$O(n * k * i)$	$O(n\sqrt{n})$

Now, in the quantum domain, a discrete-time quantum walk is used to reach the first data point in $O(\sqrt{n})$ time, the second one is $O(\sqrt{n-1})$ time, and so on.

$$O(\sqrt{n}), O(\sqrt{n-1}), O(\sqrt{n-2}), \ldots, O(\sqrt{n-(k-1)}) \leq O(k\sqrt{n}) \approx O(\sqrt{n}) \tag{7.13}$$

where k is negligible for small datasets ($k \ll n$). A detailed comparison among some state-of-the-art data clustering techniques is represented in Table 7.2, where n = total size of input.
m = no. of features available in every input data.
k = no. of clusters.
i = no. of repetition.

7.2.4 Pros and Cons

1. Unlike prior approaches (1D SCMS and 1D MCMS), our method efficiently handles outliers by linking randomly moving data points in a free space with other clusters and gradually shrinking the area in a graph.
2. To cluster the data points in quadratic time, our proposed hybrid technique uses discrete-time quantum walks, which is more realistic than clustering algorithms

that need a dedicated quantum black box to estimate the distance between pairs of points in superposition.
3. Our suggested quantum hybrid method outperforms some popular clustering techniques in terms of clustering accuracy.
4. A detailed quantum block circuit diagram is described in this chapter.
5. Because data points change their positions randomly over time and their inter-distances with the degree of each data point must be recomputed regularly, the fundamental shortcoming of our approach is determining a preset threshold value (T_θ).

7.3 Quantum Hybrid Feature Subset Selection

One of the most intriguing and significant tasks in data mining and data analysis is feature selection in data categorization [19]. It is an example of an NP-hard task, in which the complexity grows exponentially as the amount of features and the size of datasets grow. Feature selection is categorized into two types (1) wrapper-based [20, 21] and (2) filter-based [22]. The classification techniques are utilized to measure the importance of a feature reduceset in the wrapper technique, whereas classification techniques have no relevance in the second type. Therefore, when comparing the above two strategies, the first one outperforms the second one in terms of feature selection [23]. The wrapper-based technique, on the other hand, is slow when the number of input attributes is large. Numerous optimization techniques are used to achieve an efficient and reliable unsupervised feature subset selection. In this work, our suggested technique depends on the graph-theoretic idea of unified feature association map [24], which recognizes the advantages of graphical quantum theory. We first locate the reduce attributes set from the superposition of the actual candidate set by applying threshold-based Grover's search operation and amplitude amplification. This results in a quadratic speedup over the standard method. This method employs a two-stage filtering system. It examines whether the experimented dataset is acceptable for features reduction at the outset or not. The real feature selection process is then carried out using quantum parallel Grover's techniques. It will meet quickly if the experimented dataset is not acceptable for feature reduction.

7.3.1 Methodology (HQFSA)

Input: $G = (V, E, W)$ is a weighted undirected graph in which a node V defines an attribute, a connector E defines a interrelation between 2 attributes, and W signifies the strength of that link.

Output: Find a reduced features set.
Process:
Step-1: Build a feature similarity matrix using Pearson's inter-feature correlation coefficients from the given weighted undirected graph G.
Step-2: Prepare an adjacency matrix from the above feature similarity matrix after considering a prefixed threshold value $\mathrm{TH}_{\min}$.

$$A(F_i, F_j) = \begin{cases} 1, & \text{if } S(F_i, F_j) \geq \mathrm{TH}_{\min} \\ 0, & \text{otherwise} \end{cases} \tag{7.14}$$

A quantum oracle is introduced to implement the above operation on a binary matrix. As a result, oracle O works on the computational basis of the above matrix values as follows,

$$O|i\rangle|j\rangle|a\rangle = |i\rangle|j\rangle|a \oplus D_{ij}\rangle, \tag{7.15}$$

where $i \in Z_N$ and $j \in Z_M$. After taking $|a\rangle = \frac{|0\rangle-|1\rangle}{\sqrt{2}}$ (Grover's operation [23]), then a modified oracle $O^{(1)}$ is expanded as,

$$O^{(1)}|i\rangle|j\rangle\frac{|0\rangle - |1\rangle}{\sqrt{2}} = (-1)^{D_{ij}}|i\rangle|j\rangle\frac{|0\rangle - |1\rangle}{\sqrt{2}}, \tag{7.16}$$

The phase $|i\rangle|j\rangle$ is flipped when $\mathrm{sim}(F_i, F_j) \geq \mathrm{Th}_{\min}$ (i.e., $D_{ij} = 1$). To do the flipping operation, CNOT $\rightarrow \Lambda(P_x)$ (P_x is Pauli matrix) is conducted,

$$|x\rangle|y\rangle = |x\rangle|y \oplus x\rangle, \tag{7.17}$$

Therefore, a composite oracles $O^{(k)}$ that will check whether an input value which is greater than or equal to $\mathrm{TH}_{\min}$ can be prepared as,

$$O^{(k)}|i\rangle|j_1\rangle|j_2\rangle\ldots|j_k\rangle = (-1)^{\tau(i,\mathrm{sim}(F_i,F_j))}|i\rangle|j_1\rangle|j_2\rangle\ldots|j_k\rangle, \tag{7.18}$$

where $\tau(i, \mathrm{sim}(F_i, F_j)) = \prod_{l=1}^{k} D_{ijl}$ is a boolean value that indicates whether or not the input has the value $\geq \mathrm{TH}_{\min}$. If $\mathrm{sim}(F_i, F_j) \geq \mathrm{TH}_{\min}$ (i.e., $\tau(i, \mathrm{sim}(F_i, F_j)) = 1$), the phase of the state $|i\rangle|j_1\rangle|j_2\rangle\ldots|j_k\rangle$ is flipped; otherwise, the phase is unaffected.

The sequential steps of the operation is explained as,

A. Prepare the starting states for the initial registers, $|i\rangle|j\rangle|0\rangle\frac{|0\rangle-|1\rangle}{\sqrt{2}}$.
B. Apply $O^{(1)}$ on the state, $|i\rangle|j\rangle|D_{ij}\rangle\frac{|0\rangle-|1\rangle}{\sqrt{2}}$.
C. Perform the operation $\Lambda(P_x)$ to the $\frac{|0\rangle-|1\rangle}{\sqrt{2}}$ qubit, and we have the modified state as, $(-1)^{D_{ij}\geq \mathrm{Th}_{\min}}|i\rangle|j\rangle|D_{ij}\rangle\frac{|0\rangle-|1\rangle}{\sqrt{2}}$.
D. Reverse the step B (returning the registers to their starting state) and discarding the last $\frac{|0\rangle-|1\rangle}{\sqrt{2}}$ qubit, followed by the Oracle $O^{(k)}$ implementation (Fig. 7.2).

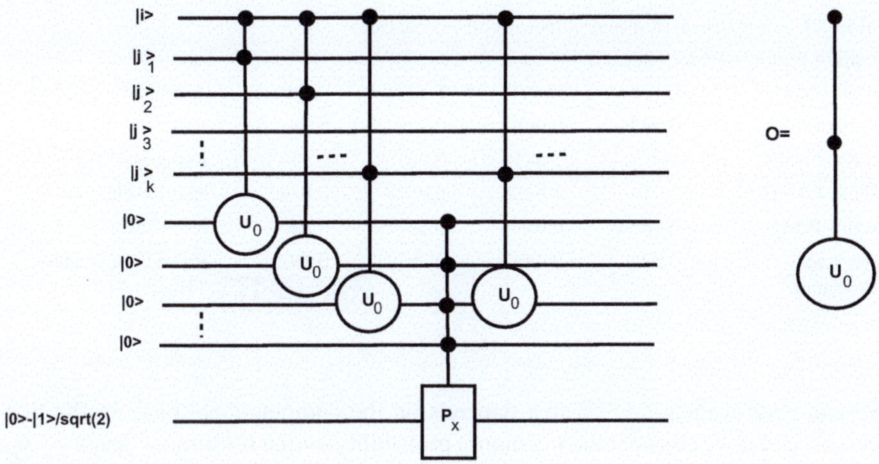

Fig. 7.2 Oracle $O^{(1)}$ circuit diagram with O and CNOT$\Lambda(P_x)$

Step-3: We have applied a direct parallel quantum amplitude estimation(QAE) to find the maximum correlated values ($V_c^{(1)}$) in a quantum environment. A related Oracle $O_j^{(k)}$ is used to do this operation.

$$O_j^{(k)}|i\rangle = (-1)^{D_{ij}}|i\rangle, \quad \text{where, } D_{ij} \geq \mathrm{TH}_{\min} \tag{7.19}$$

The related oracle uses a Grover operator, that is defined as follows,

$$G_j^{(k)}|i\rangle = (2|X_N\rangle\langle X_N| - I_N)|O_j^{(k)}\rangle, \tag{7.20}$$

where $|X_N\rangle = \frac{1}{\sqrt{N}}\sum_{i=0}^{N-1}|i\rangle$ and I_N is the identity matrix with dimension N. The $G_j^{(k)}$ has 2 eigen values, $\lambda_{\pm} = \mathrm{e}^{\pm 2i\theta_j^{(k)}}$ $(i = \sqrt{-1})$, and the eigenvectors are $|\phi_{j\pm}^{(k)}\rangle$. Now, the begining states of the 2 registers are $(\frac{\sum_{t=0}^{T-1}|t\rangle}{\sqrt{T}})|X_N\rangle$, where $T = 2^t$, $|t\rangle$ are the basic states of the 1st register and $N \geq \mathrm{TH}_{\min}$, after ignoring the global phase we receive,

$$|\phi_j^{(k)}\rangle = \frac{\mathrm{e}^{i}\theta_j^{(k)}}{\sqrt{2}}\left|\varepsilon_T\left(\frac{\theta_j^{(k)}}{\pi}\right)\right\rangle|\phi_{j+}^{(k)}\rangle - \frac{\mathrm{e}^{-i}\theta_j^{(k)}}{\sqrt{2}}\left|1 - \varepsilon_T\left(\frac{\theta_j^{(k)}}{\pi}\right)\right\rangle|\phi_{j-}^{(k)}\rangle, \tag{7.21}$$

However, $\varepsilon_T(\omega) = |T\omega\rangle$ when $T\omega$ is an integer, otherwise,

$$|\varepsilon_T(\omega)\rangle = \sum_{y=0}^{T-1}\frac{\mathrm{e}^{2\pi i(T\omega - y)} - 1}{T\left(\mathrm{e}^{\frac{2\pi i(T\omega - y)}{T}} - 1\right)}|y\rangle \tag{7.22}$$

Table 7.3 Comparison of different metrics on iris dataset

Support vector machine (linear)					
Algorithms	Precision	Recall	F-score	Accuracy	Attributes-selection
Actual	0.947	0.968	0.979	0.987	Full
Score-Fisher	0.97	0.945	0.96	0.959	Reduceset(2)
Score-OVAF	0.97	0.946	0.957	0.958	Reduceset(2)
Score-FDAF	0.96	0.95	0.96	0.955	Reduceset(2)
Proposed algorithm	0.98	0.989	0.991	0.992	Reduceset(3) and $\text{TH}_{\text{min}} = 0.7$

We can now evaluate $|\phi j^{(k)}\rangle$ that depends on the computational basis in the first register, and if we consider the maximum probability output for some y'_j and $T - y'_j$, then $\sin^2\left(\frac{\pi y'_j}{T}\right) = \sin^2\left(\frac{\pi(T-y'_j)}{T}\right)$. As a result, $\sin^2\left(\frac{\pi y'_j}{T}\right)$ or $\sin^2\left(\frac{\pi(T-y'_j)}{T}\right)$ is used as an estimate of $\sin^2(\theta_j^{(k)})$. Due to the linearity of the unitary operator, we will acquire the last state for $n \times n$ linked values. The first register is used to hold the output values parallely.

Step-4: We execute parallel amplitude amplification technique on the first register of $|\phi j^{(k)}\rangle$ to look for the values such as $\sin^2\left(\frac{\pi y'}{T}\right) \geq \text{TH}_{\text{min}}$ or $\sin^2\left(\frac{\pi(T-y')}{T}\right) \geq \text{TH}_{\text{min}}$ (see Eq. 7.21). Finally, out of 2^n configurations, count all feasible configurations with a similarity measure greater than a threshold.

Step-5: To find all the values ($N_c = 2^n$) that are bigger than TH_{min}, we will evaluate the first and third registers for $(O(N_c^{(k)}))$ times. However, this evaluation can be performed multiple times to obtain the best measurement in a quantum system independently of quantum noise.

7.3.2 *Result Analysis*

The entire experiment is conducted on a classical system due to the infrastructure scarcity of quantum computer. Like the previous case, the same configuration has been used here. In this chapter, the proposed algorithm is first implemented to some widely known public datasets to acquire the reduce feature subset, after which the learning model's precision, recall, F-score, and accuracy are measured, and finally, they are compared with some popular algorithms. For the experimented datasets under consideration, the proposed approach has the maximum average accuracy value (Tables 7.3, 7.4, 7.5, 7.6, 7.7, 7.8, and 7.9).

Table 7.4 Comparison of different metrics on Wisconsin breast cancer dataset

SVM (linear)

Algorithms	Precision	Recall	F-score	Accuracy	Attributes-selection
Actual	0.939	0.946	0.952	0.949	Full
Correlation-based FS	0.859	0.88	0.862	0.869	Reduceset
Minimum redundancy FS	0.929	0.947	0.934	0.938	Reduceset
Discrete sine-cosine FS	0.925	0.927	0.921	0.929	Reduceset
Proposed algorithm	0.949	0.938	0.947	0.949	Reduceset and $TH_{min} = 0.69$

Table 7.5 Comparison of different metrics on vehicle dataset

Support vector machine

Algorithms	Precision	Recall	F-score	Accuracy	Attributes-selection
Actual	0.779	0.79	0.778	0.779	Full eightfold crossvalidation
Correlation-based	0.278	0.29	0.279	0.289	Reduceset
Minimum redundancy	0.46	0.458	0.452	0.459	Reduceset
Discrete sine-cosine	0.48	0.49	0.485	0.479	Reduceset
Proposed algorithm	0.789	0.769	0.78	0.787	Reduceset(11) and eightfold CV $TH_{min} = 0.68$
Proposed algorithm	0.83	0.833	0.82	0.832	Reduceset(11) and $TH_{min} = 0.68$

Table 7.6 Comparison of different metrics on glass dataset

Support vector machine

Algorithms	Precision	Recall	F-score	Accuracy	Attributes-selection
Actual	0.66	0.648	0.66	0.659	Full (not CV)
Correlation-based	0.475	0.492	0.48	0.487	Reduceset
Minimum redundancy	0.50	0.519	0.505	0.509	Reduceset
Discrete sine-cosine	0.63	0.629	0.631	0.629	Reduceset
Score-Fisher	0.68	0.70	0.708	0.689	Reduceset(5)
Score-OVAF	0.649	0.69	0.69	0.679	Reduceset(4)
FDAF score	0.70	0.712	0.698	0.71	Reduceset(5)
Proposed algorithm	0.768	0.749	0.749	0.895	Reduceset(6) and eightfold CV $TH_{min} = 0.57$
Proposed algorithm	0.668	0.672	0.669	0.669	Reduceset(4) and $TH_{min} = 0.69$

Table 7.7 Comparison of different metrics on sonar dataset

Nearest neighbors ($K = 3$)					
Algorithms	Precision	Recall	F-score	Accuracy	Attributes-selection
Actual	0.761	0.762	0.749	0.76	Full
Correlation-based	0.539	0.55	0.569	0.568	Reduceset(40)
Minimum redundancy	0.559	0.585	0.579	0.569	Reduceset(40)
Discrete sine-cosine	0.69	0.686	0.72	0.72	Reduceset(40)
Proposed algorithm	0.79	0.78	0.792	0.739	Reduceset(30) and $TH_{min} = 0.67$

Table 7.8 Comparison of different metrics on wine dataset

Nearest neighbors ($K = 3$)					
Algorithms	Precision	Recall	F-score	Accuracy	Attributes-selection
Actual	0.88	0.849	0.848	0.849	Full
Correlation-based	0.719	0.728	0.71	0.714	Reduceset
Minimum redundancy	0.829	0.82	0.83	0.829	Reduceset
Discrete sine-cosine	0.859	0.86	0.856	0.859	Reduceset
Score-Fisher	0.857	0.867	0.865	0.869	Reduceset(7)
Score-OVAF	0.848	0.87	0.85	0.857	Reduceset(7)
Proposed algorithm	0.88	0.868	0.865	0.869	Reduceset(8) $TH_{min} = 0.69$

Table 7.9 Comparison in terms of accuracy

Databases	Accuracy with original attributes	Accuracy with reducedset (HQFSA)	Classification algorithms	Win/loss
Iris	0.968	0.987	Support vector machine (linear)	Win
Wine	0.847	0.869	Nearest neighbors ($k = 3$)	Win
Vehicle	0.792	0.82	Support vector machine (linear)	Win
Glass	0.678	0.89	Support vector machine (linear)	Win
Sonar	0.749	0.735	Nearest neighbors ($k = 3$)	Loss
Breast cancer	0.948	0.947	Support vector machine (linear)	Tie
Ionosphere	0.894	0.923	Support vector machine (linear)	Win

Table 7.10 Comparison on mean execution time (seconds)

Databases	WOAfs [25]	PeSOAfs [26]	ABCfs [27]	Proposed algorithm
Sonar	1.31	0.992	1.89	0.24
Wine	0.9	0.925	1.88	0.5
Iris	0.928	0.82	1.98	0.23
Glass	1.19	0.9	2.1	0.46
Vehicle	0.89	1.72	1.5	0.49
Breast cancer	1.4	1.5	1.39	0.69
Ionosphere	1.2	1.4	1.96	0.79

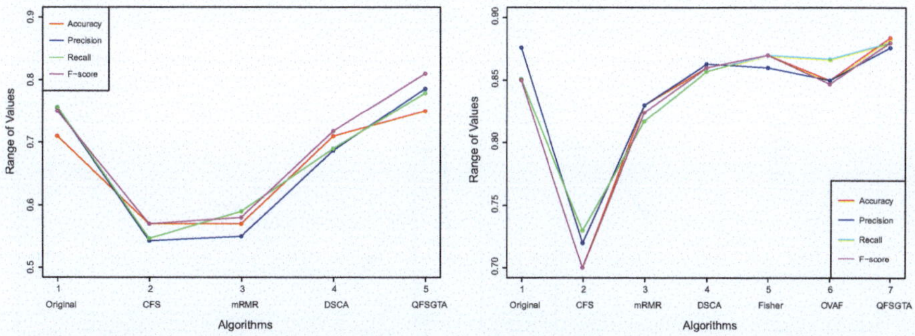

(a) Comparison of performance of algorithms for sonar dataset

(b) Comparison of performance of algorithms for wine dataset

Fig. 7.3 Comparison among different algorithms in terms of various measures using *K*NN classifier

7.3.3 Complexity Analysis

Table 7.10 shows the computational time analysis in terms of average values over the 30 runs. The computing time, on the other hand, is increased with the sample size. For datasets with a modest number of features or samples, such as glass, iris, and others, it takes only a few minutes. Two comparison barchart plots (Figs. 7.3 and 7.4) on iris and wine datasets are shown as a demo using *K*NN and SVM classifiers, respectively. Figure 7.5 shows the summary of performance in terms of some statistical measures. Finally, Fig. 7.6 shows an overall win/loss summary of the comparison result (Table 7.11).

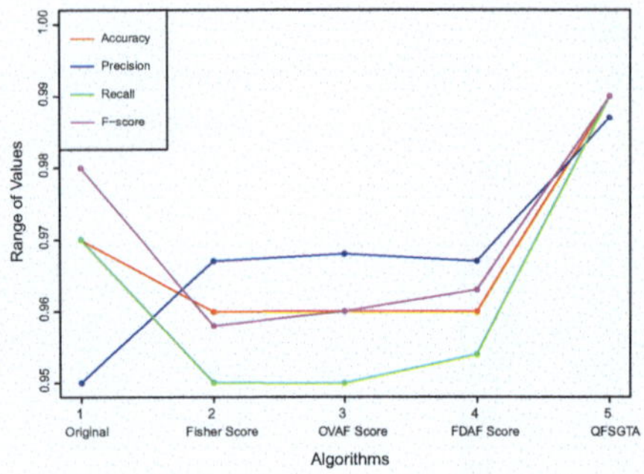

(a) Comparison of performance of algorithms for iris dataset

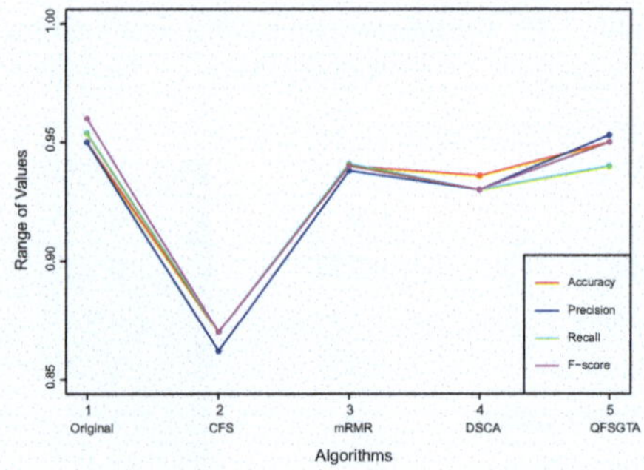

(b) Comparison of performance of algorithms on Wisconsin Breast cancer dataset

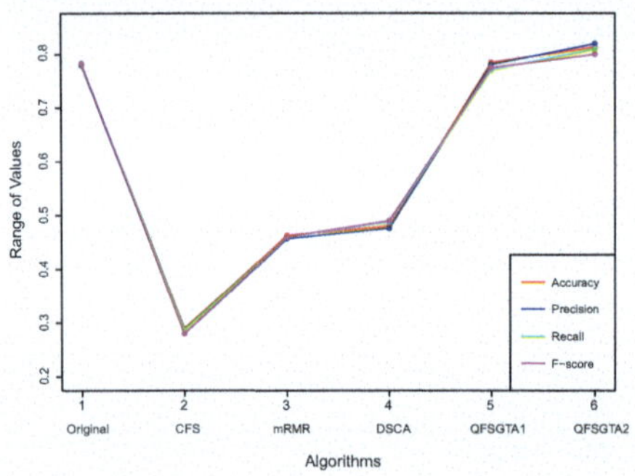

(c) Comparison of performance of algorithms on vehicle dataset

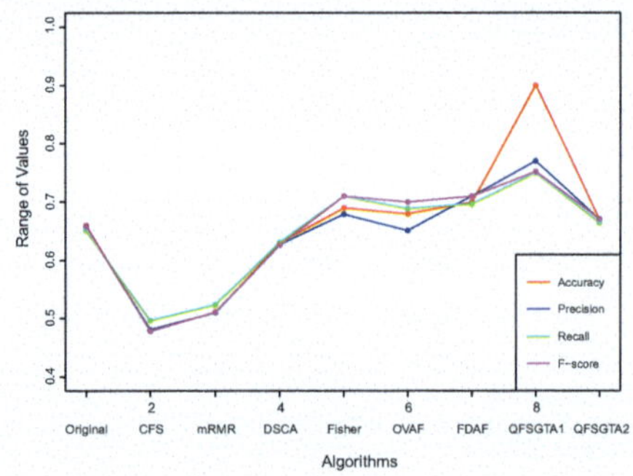

(d) Comparison of performance of algorithms on glass dataset

Fig. 7.4 Comparison among different algorithms in terms of various measures using SVM-linear classifier

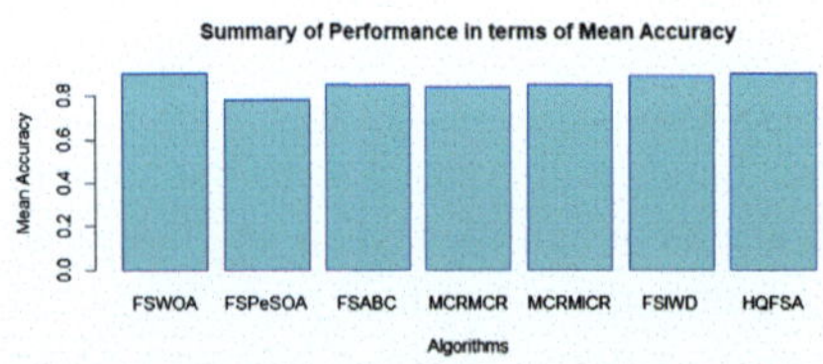

(a)

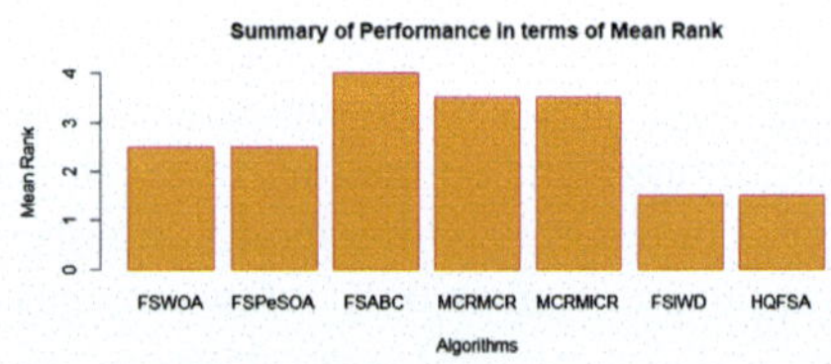

(b)

Fig. 7.5 Results summary

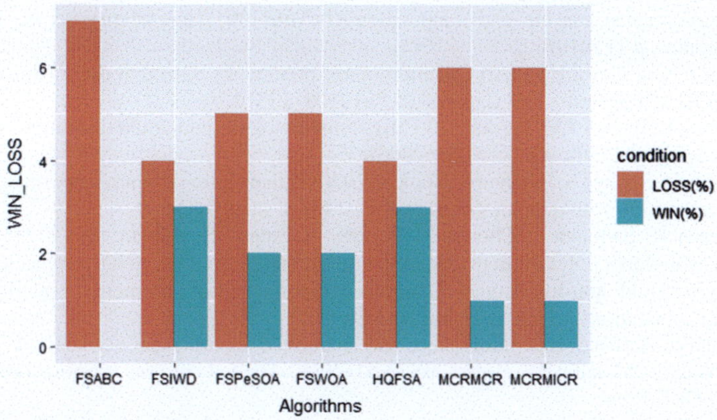

Fig. 7.6 WIN versus LOSS of results

Table 7.11 Results summary

Algorithms	Avg. accuracy	Avg. rank	No. of win/tie
WOAfs	0.89	2.48	Two
PeSOAfs	0.779	2.49	Two
ABCfs	0.848	3.95	Zero
MCR-MCR	0.839	3.45	One
MCR-MICR	0.849	3.46	One
IWDfs	0.887	1.49	Three
HQFSA	0.912	1.48	Three

7.4 Conclusion

These two quantum hybrid approaches described in this chapter give a different essence of how quantum principles can be used for doing better data clustering and feature subset selection than some well-known classical counterparts. It shows the major comparisons in terms of accuracy and speedup factors. These two works represent the doorsteps of quantum-classical hybrid techniques suitable in the NISQ era.

References

1. Nielsen MA, Chuang IL (2010) Quantum computation and quantum information. Cambridge University Press, New York, Hardback. ISBN 978-1-107-00217-3
2. Sleeman J, Dorband J, Halem M (2020) A hybrid quantum enabled RBM advantage: convolutional autoencoders for quantum image compression and generative learning. In: Quantum

information science, sensing, and computation XII, vol 11391. International Society for Optics and Photonics, p 113910B
3. Xu P, He Z, Qiu T, Ma H (2020) Quantum image processing algorithm using edge extraction based on Kirsch operator. Opt Express 28(9):12508–12517
4. Yan F, Chen K, Iliyasu AM, Hirota K (2020) Circuit-based modular implementation of quantum ghost imaging. IEEE Access 8:23054–23068
5. Henderson M, Shakya S, Pradhan S, Cook T (2020) Quanvolutional neural networks: powering image recognition with quantum circuits. Quant Mach Intell 2(1):1–9
6. Jagatheesan K, Samanta S, Choudhury A, Dey N, Anand B, Ashour AS (2018) Quantum inspired evolutionary algorithm in load frequency control of multi-area interconnected thermal power system with non-linearity. In: Quantum computing: an environment for intelligent large scale real application. Springer, Cham, pp 389–417
7. Kak S (2000) Active agents, intelligence and quantum computing. Inf Sci 128:1–17
8. Dey N, Wagh S, Mahalle PN, Pathan MS (eds) (2019) Applied machine learning for smart data analysis. CRC Press
9. Ezhov AA (2001) Pattern recognition with quantum neural networks. In: Proceedings of advances in pattern recognition, pp 60–71
10. Ventura D (2002) Pattern classification using a quantum system. In: Proceedings of the 6th joint conference on information science, pp 537–540
11. Li Q, He Y, Jiang JP (2011) A hybrid classical-quantum clustering algorithm based on quantum walks. Quant Inf Process 10(1):13–26
12. Trugenberger CA (2002) Quantum pattern recognition. arXiv:quant-ph/0210176v2
13. Lu S, Braunstein SL (2013) Quantum decision tree classifier. Quant Inf Process
14. Harrow AW, Hassidim A, Lloyd S (2009) Quantum algorithm for linear systems of equations. Phys Rev Lett 103(15):150502
15. Zhang Y, Ni Q (2020) Recent advances in quantum machine learning. Quant Eng 2(1):e34
16. Portugal R (2013) Quantum walks and search algorithms. In: Quantum science and technology, Springer Science and Business Media
17. Philipp P, Portugal R (2017) Exact simulation of coined quantum walks with the continuous-time model. Quant Inf Process 16(1)
18. Soliman OS, Rassem A (2012) Correlation based feature selection using quantum bio inspired estimation of distribution algorithm. In: International workshop on multi-disciplinary trends in artificial intelligence. Springer, Berlin, pp 318–329
19. Abualigah LM, Khader AT, Hanandeh ES (2018) A new feature selection method to improve the document clustering using particle swarm optimization algorithm. J Comput Sci 25:456–466
20. Yang C-H, Chuang L-Y, Yang C-H (2010) IG-GA: a hybrid filter/wrapper method for feature selection of microarray data. J Med Biolog Eng 30(1):23–28
21. Kohavi R, John GH (1997) Wrappers for feature subset selection. Artif Intell 97(1–2):273–324
22. Goswami S, Chakraborty S, Guha P, Tarafdar A, Kedia A (2019) Filter-based feature selection methods using Hill climbing approach. In: Natural computing for unsupervised learning. Springer, Cham, pp 213–234
23. Song G, Jang K, Kim H, Eum S, Sim M, Kim H, Lee W, Seo H (2022) SPEEDY quantum circuit for Grover's algorithm. Appl Sci 12(14):6870
24. Das AK, Goswami S, Chakrabarti A, Chakraborty B (2017) A new hybrid feature selection approach using feature association map for supervised and unsupervised classification. Exp Syst Appl 88:81–94
25. Mafarja M, Mirjalili S (2018) Whale optimization approaches for wrapper feature selection. Appl Soft Comput 62:441–453
26. Dasgupta A, Banerjee A, Dastider AG, Barman A, Chakraborty S (2019) A study and analysis of a feature subset selection technique using Penguin search optimization algorithm (PeSOA-FS). In: Data science: theory, analysis, and applications. CRC Press, Taylor and Francis. arXiv preprint arXiv:1611.09347
27. Hancer E, Xue B, Zhang M, Karaboga D, Akay B (2018) Pareto front feature selection based on artificial bee colony optimization. Inf Sci 422:462–479

Chapter 8
Quantum Computing in Image Processing

8.1 Introduction

In current years, quantum image processing becomes one of the popular research topics along with the quantum machine learning. Like quantum machine learning, the similar kind of challenges are facing while handling large matrices in the image processing operations (image denoising, image segmentation [1], image compression, image security, etc.). Quantum algorithms make a substantial contribution to the field of quantum image processing by utilizing entanglement, which is perhaps the most important resource of many-qubit quantum systems. It could pave the way for more effective techniques of representing and acquiring information about the objects in quantum images. However, quantum principles not only explore the benefits in the field of machine learning but also show how image denoising, image edge detection, morphological operation can be performed more efficient way (improved image quality and experimental speed up) using the fast quantum Fourier and wavelet transform techniques in a binary quantum system. Quantum computing combines its principles with image processing to represent effective information processing technology by taking advantage of quantum computation.

8.2 Quantum Image Denoising

The basic goal of denoising is to eliminate noise while maintaining as much of the vital information as feasible. Numerous researches on removing noise from images have been conducted using different kind of transforms in the past [2–5]. However, no work on image denoising in the quantum domain has been done too far. In addition, when encoding the desired operation on a quantum computer, a unitary transformation must be linear and reversible. Irreversibility can be overcome by summing "ancilla" qubits that allow us to memorize the input that corresponds to each output. Nonlinear transitions, on the other hand, are still an issue in the quantum

S. Chakraborty and L. Dey, *Computing for Data Analysis: Theory and Practices*, Data-Intensive Research, https://doi.org/10.1007/978-981-19-8004-6_8

domain. Fortunately, unitary operators may be used to represent Fourier transformations, Hadamard–Walsh transforms, and various other transforms. This is why the quantum computing community is currently studying quantum Fourier transforms (QFT) and quantum wavelet transforms (qwt). A handful scientists are concentrating on few quantum noise removal algorithms to tackle with noise difficulties [6–8]. The work proposes and implements a quantum image filtering model based on QFT [6, 7]. In paper [8], the filtering circuit is built using a black box quantum oracle. The inner black box circuit is hidden, which is a major disadvantage of this technology. However, in our suggested approach, the detail quantum circuit design has been depicted in full.

We concentrate on image denoising process using wavelet transform in quantum domain, inspired by the previous quantum filtering approaches. We employ the Daubechies fourth-order wavelet kernel idea and apply a specialized parallel quantum oracle to find the required wavelet coefficients, in addition to the elementary quantum addition operation [9]. Our work's primary strength is a discussion of a quantum technique for denoising grayscale images utilizing the quantum wavelet transform along with significant quality measure parameters. The comprehensive architecture of all quantum circuits for removing noise process of images utilizing qwt is also presented in this chapter. Finally, this chapter provides a detailed analysis of results of the suggested scheme applying on a variety of common image samples, as well as a thorough comparison on various image quality measurement parameters, computation of time, and storage size.

8.2.1 Background

The primary purpose of this section is to discuss the quantum Daubechies fourth-order wavelet transform. For image processing, we use Daubechies fourth-order wavelet kernel of dimension $2n$ in our research [9]. The factorization process of the transform matrix is given below,

$$D_{2^n}^{(4)} = (I_{2^{n-1}} \otimes C_1) Q_{2^n} (I_{2^{n-1}} \otimes C_0') \tag{8.1}$$

wherein, C_1 and C_0' are 2 unitary matrices that are expressed as,

$$C_0' = 2 \begin{bmatrix} C_3 & -C_2 \\ C_2 & C_3 \end{bmatrix}$$

and

$$C_1 = \frac{1}{2} \begin{bmatrix} \frac{C_0}{C_3} & 1 \\ 1 & \frac{C_1}{C_2} \end{bmatrix}$$

where, $C_0 = (3+\sqrt{3})/4\sqrt{2}$, $C_1 = (3-\sqrt{3})/4\sqrt{2}$, $C_2 = (1-\sqrt{3})/4\sqrt{2}$, $C_3 = (1+\sqrt{3})/4\sqrt{2}$. $C_0' = N(C_0)$

$$N = \begin{bmatrix} 0 & 1 \\ 1 & 0 \end{bmatrix}$$

and Q_{2^n} is shown below,

$$Q_{2^n} = \begin{bmatrix} 0\,1 & & & & & & \\ 0\,0\,1 & & & & & & \\ 0\,0\,0\,1 & & & & & & \\ 0\,0\,0\,0\,1 & & & & & & \\ \cdot\;\cdot\;\cdot\;\cdot\;\cdot\;\cdot\;\cdot & & & & & & \\ \cdot\;\cdot\;\cdot\;\cdot\;\cdot\;\cdot\;\cdot & & & & & & \\ 0\,0\;.\;.\;.\;.\;0\,0\,0 & & & & & & \\ 1\,0\;.\;.\;.\;.\;0\,0\,0 & & & & & & \end{bmatrix}$$

However, it can be factored as,

$$Q_{2^n} = F_{2^n} P_{2^n} T_{2^n} P_{2^n} F_{2^n}^* \tag{8.2}$$

where, F_{2^n} represents Cooley–Tukey FFT factorization for a $2n$-dimensional vector and $F_{2^n}^*$ is the conjugate transpose of F_{2^n}, P_{2^n} is a bit-reversal permutation matrix. It can be represented as,

$$P_{2^n} : |a_{n-1}a_{n-2}\cdots a_1 a_0\rangle \rightarrow |a_0 a_1 \cdots a_{n-2}a_{n-1}\rangle, \tag{8.3}$$

and T_{2^n} is a diagonal matrix given as, $T_{2^n} = \text{diag}[1, \omega_{2^n}, \omega_{2^n}^2, \ldots, \omega_{2^n}^{2^n-1}]$, $\omega_{2^n}^2 = e^{\frac{-2i\pi}{2^n}}$, $i = \sqrt{-1}$. However, the final product matrix $P_{2^n} T_{2^n} P_{2^n}$ can be factorized into,

$$P_{2^n} T_{2^n} P_{2^n} = (I_{2^{n-1}} \otimes G(\omega_{2^n}^{2^{n-1}}))\cdots(I_{2^{n-i}} \otimes G(\omega_{2^n}^{2^{n-i}}) \otimes I_{2^{i-1}})\cdots(G(\omega_{2^n}) \otimes I_{2^{n-1}}), \tag{8.4}$$

wherein, $G(\omega_{2^n}^k) = \text{diag}(1, \omega_{2^n}^k)$, $k \in (0, n-1)$ and,

$$G(\omega_{2^n}^k) = \begin{bmatrix} 1 & 0 \\ 0 & \omega_{2^n}^k \end{bmatrix}$$

8.2.2 Methodology

Procedure

Step 1. In the classical domain, we must first combine the actual image with a noisy image. To describe an image and create a noisy image in the quantum domain, we adopt the well-known FRQI [10] technique. Using the FRQI approach, we create a quantum register to hold a $2^n \times 2^n$ quantum image $|Q\rangle$.

$$|Q\rangle = \sum_{i=0}^{2^{2n-1}} (|c_i\rangle) \otimes |i\rangle \tag{8.5}$$

$$|Q\rangle = \sum_{i=0}^{2^{2n-1}} (|c_i\rangle) \otimes |y_i\rangle|x_i\rangle,\ |y_i\rangle = |y_{n-1}y_{n-2}\cdots y_0\rangle \text{ and } |x_i\rangle = |x_{n-1}x_{n-2}\cdots x_0\rangle, \tag{8.6}$$

whereas, $|y_i\rangle|x_i\rangle \in [0, 1], i = 0, 1, 2, \ldots, n$. Therefore, we generate a noisy image $|N\rangle$ with the help of the similar FRQI technique.

$$|N\rangle = \sum_{i=0}^{2^{2n-1}} (|c_i\rangle) \otimes |i\rangle \tag{8.7}$$

To convert the above input image into its matching noisy image, we'll use P_k transform, i.e., unitary in nature. However, in order to generate a noisy image, we require a predetermined pattern-based quantum phase change operation, because during the retrieval of the noisy image using Iqwt, a same kind of pattern need to be observed.

$$P|N\rangle = \left(\prod_{k=0}^{2^{2n-1}} P_k\right)(|N\rangle) = |PN\rangle \tag{8.8}$$

whereas, P_k is the transformation that can be defined as,

$$P_k = \left(I \otimes \sum_{j=0, j\neq i}^{2^{2n-1}} |j\rangle|j\rangle\right) + R(2\phi_k) \otimes |k\rangle|k\rangle \tag{8.9}$$

whereas,

$$I = \begin{vmatrix} 1 & 0 \\ 0 & 1 \end{vmatrix}$$

and

$$R(2\phi_k) = \begin{vmatrix} \cos\phi_k & -\sin\phi_k \\ \sin\phi_k & \cos\phi_k \end{vmatrix}$$

whereas, $R(2\phi_k)$ = operation with rotation. Therefore, we can forward further as,

$$P_k(|N\rangle) = \frac{1}{2^n} \sum_{i=0}^{2^{2n-1}} (R(2\phi_k) \otimes I^{2n})(\cos\theta_i|0\rangle + \sin\theta_i|1\rangle) \otimes |i\rangle) \tag{8.10}$$

$$P_k(|N\rangle) = \frac{1}{2^n} \sum_{i=0, i \neq k}^{2^{2n-1}} (\cos\theta_i|0\rangle + \sin\theta_i|1\rangle) \otimes |i\rangle) + (\cos(\theta_i + \phi_k)|0\rangle + \sin(\theta_i + \phi_k)|1\rangle) \otimes |k\rangle \tag{8.11}$$

We get our initial scrambled image after running these set of operations 2^{2n} number of times.

Step 2. On the images above, apply a multilevel discrete wavelet decomposition.

$$\begin{aligned} \text{qwt}(|Q\rangle) &= \text{qwt}\left(\frac{1}{2^n}\sum_{i=0}^{2^{2n-1}}(\cos\theta_i|0\rangle + \sin\theta_i|1\rangle) \otimes |i\rangle\right) \\ &= \frac{1}{2^n}\sum_{i=0}^{2^{2n-1}} \text{qwt}(\cos\theta_i|0\rangle + \sin\theta_i|1\rangle \otimes |i\rangle) \\ &= \sum_{i=0}^{2^{2n-1}} |wc_i\rangle \otimes |i\rangle \end{aligned} \tag{8.12}$$

Step 3. Now, add the basic quantum image with its noisy counterpart. The wavelet coefficients qwt($|Q\rangle$) are integrated in the processed quantum noisy image $|PN\rangle$, and the output is shown below.

$$\text{qwt}(|QN\rangle) = \text{qwt}(|Q\rangle) + |P_k N\rangle) = \sum_{i=0}^{2^{2n-1}} |\text{wmc}_i\rangle \otimes |i\rangle \tag{8.13}$$

qwt($|Q\rangle$) denotes the original image's qwt qubits, while qwt($|QN\rangle$) denotes the qubits after incorporating the noisy image. This method of embedding is totally reversible. The entire procedure must adhere to the quantum mechanics premise. This embedding technique is implemented using a simple quantum adder network, as shown in Fig. 8.1. To store the above both type of images, two quantum registers are required at first. The original image is then subjected to a qwt operation, as well as a succession of P_k operations in the embedding process. Figure 8.2 [11] depicts the fundamental quantum sum and carry network that aids in the construction of the quantum adder circuit.

Step 4. Select an acceptable threshold range using a thresholding approach. A quantum black box (U_f) [9] is implemented and also leverage the notion of parallelism toward quantum computing to perform the threshold checking procedure concurrently in order to accomplish the thresholding operation in the frequency domain on each pixels of the given image. The output image of high pass or low pass filtering is the summation of the images containing maximum frequencies and the images having the respective minimum frequencies. We must adopt an operation based on threshold value to distinguish minimum and maximum coefficient components from

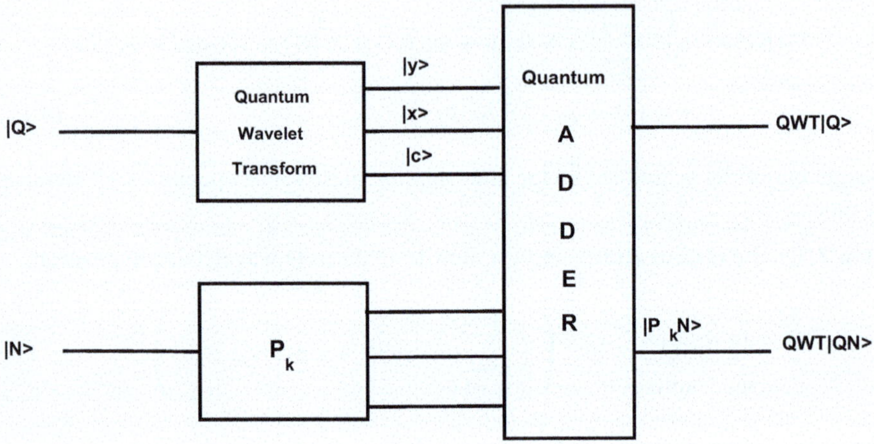

Fig. 8.1 Circuit of quantum noisy image addition process

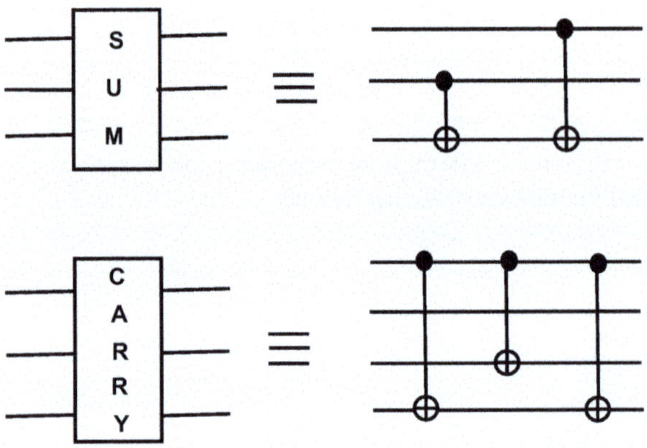

Fig. 8.2 Sum and carry adder quantum circuit

the previous phase's final product. To distinguish between high and low frequencies images, qwt($|QN\rangle$) and the added qubit $|0\rangle$ is employed.

A thresholding function $f : (0, 1, 2, \ldots, 2^{m+2n} - 1) \rightarrow (0, 1)$ is embedded in a oracle operator U_f on the state qwt$|QN\rangle \otimes |0\rangle$ (refer Eq. 8.14). The $|QN_{\text{out}}\rangle$ state is defined as,

$$\begin{aligned} |QN_{\text{out}}\rangle &= |H_{cf}\rangle^{\otimes m} |L_{cf}\rangle^{\otimes m} |TH\rangle_m \text{qwt}|QN\rangle_{m+2n}|0\rangle \\ &= |0\rangle^{\otimes m} U_f(|\text{TH}\rangle_m \text{qwt}|QN\rangle_{m+2n}|0\rangle) \\ &= |0\rangle^{\otimes m}(\text{qwt}|QN^{Hf}\rangle + \text{qwt}|QN^{Lf}\rangle|0\rangle) \end{aligned} \tag{8.14}$$

The thresholding function is being used in the quantum circuit of Fig. 8.3 to hold the coefficients states which are greater than the threshold value (denoting maximum frequency (H_{cf}) components) and evaluate it on the ancillary and controlled swap to

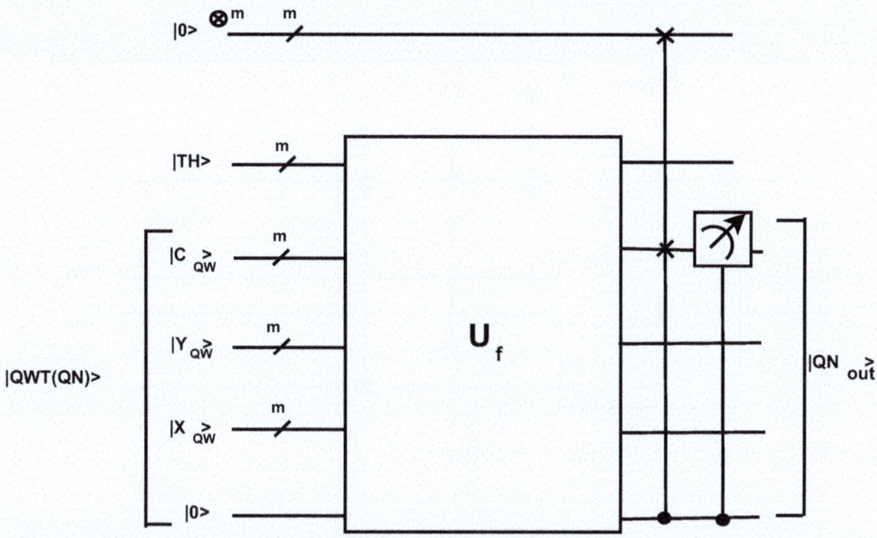

Fig. 8.3 A quantum thresholding wavelet coefficients-based circuit

$|0\rangle$ (denoting minimum frequency (L_{cf}) elements) if it is smaller than the threshold. The threshold value is stored in $|\mathrm{TH}\rangle$, a m bit register. The oracle operator U_f is implemented with the help of a special comparator work on qubit string in unitary fashion. The working of qubit string comparator, U_{CMP}, is characterized as follows,

$$U_{\mathrm{CMP}}|a\rangle|b\rangle|0\rangle^{\otimes p}|0\rangle|0\rangle = |a\rangle|b\rangle|\psi\rangle|o_1\rangle|o_2\rangle \tag{8.15}$$

It utilizes quantum registers $|a\rangle$ with p qubits and $|b\rangle$ maintain the states of compare, ancillae is represented by $p + 2$ and $|o_1\rangle$, $|o_2\rangle$ keep the comparison results.

$$\begin{array}{ll} |o_1\rangle = |o_2\rangle = 0, & \text{if } a = b \\ |o_1\rangle = 1, |o_2\rangle = 0, & \text{if } a > b \\ |o_1\rangle = 0, |o_2\rangle = 1, & \text{if } a < b \end{array} \tag{8.16}$$

The quantum circuit that handles this comparison operator, as well as our oracle operator based on it, is depicted in Fig. 8.4. If $c \geq \mathrm{TH}$, the oracle qubit is flipped by this oracle operator U_f, where $\mathrm{TH} = |\mathrm{TH}_1\rangle|\mathrm{TH}_0\rangle$ is the threshold and $|C\rangle = |C_1\rangle|C_0\rangle$ is the updated coefficients.

With the size of the color space, the expense of constructing suggested quantum black box rises straightly. It is worth noting that the size of the input image has no bearing.

Step 5. Inverse qwt and inverse P_k^{-1} operations are applied to fetch the quantum image states of noise from the combined form of image.

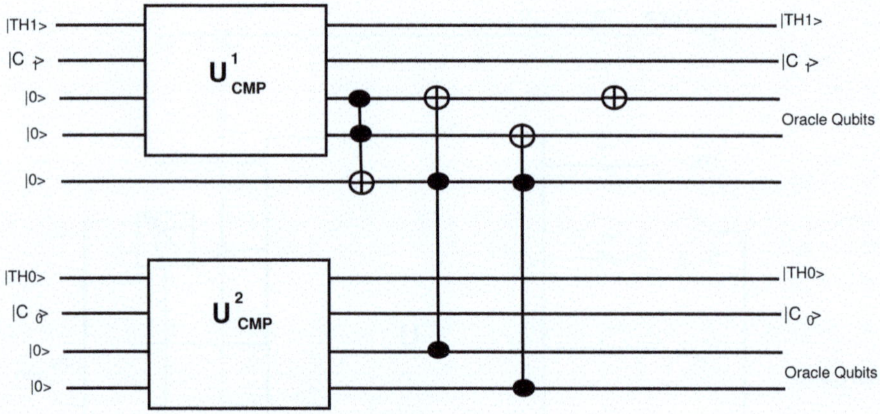

Fig. 8.4 A quantum circuit for qubits comparison

In the extraction stage, we use qwt^{-1} and the P_k^{-1} to retrieve the noisy image $|N\rangle$ from the integrated component $\text{qwt}|QN\rangle$. The embedding technique is totally reversible since all of the transforms employed in quantum computation are unitarily defined by matrices.

$$\begin{aligned}\text{Inverse qwt}[\text{qwt}(|QN_{\text{out}}\rangle)] &= \text{qwt}^{-1}[\text{qwt}(|Q\rangle) + |P_k N\rangle] \\ &= |Q\rangle + \text{qwt}^{-1}(|P_k N\rangle)\end{aligned} \tag{8.17}$$

Therefore, the ultimate noisy image can be extracted as,

$$|N\rangle = P_k^{-1}[\text{qwt}(|QN\rangle) + \text{qwt}(|Q\rangle)] \tag{8.18}$$

whereas, P_k^{-1} is the inverse transform.

Image denoising, on the other hand, is generally a preprocessing step in most image processing scenarios, and the result is frequently used as an input for numerous further processes. As a result, the extraction phase is not required. It is analyzed again, and extraction is required at the end of the quantum phase.

8.2.3 Analysis of Results

In this section, we have gone through the visual quality, complexity in terms of time and storage space in detail. We analyze multiple PSNR (peak signal to noise ratio), QIFM (quantum image fidelity metric), and MSE (mean squared error) metrics to determine the image quality. The distinguish between the actual image and its noisy counterpart is defined by PSNR. Instead of the scrambled quantum image described before, we use two relatively common noises (gaussian and salt and pepper) to

Table 8.1 Comparison among quantum fidelity metric values

Types of images	Noisy QIFM (%)	Proposed scheme QIFM (%)	Classical case (%)
Lena (gaussian)	74.02	85.98	85.51
Lena (salt–pepper)	95.05	96.91	97.25
Baboon (gaussian)	75.00	90.23	89.33
Baboon (salt–pepper)	95.08	98.41	98.91
Cameraman (gaussian)	77.11	88.96	87.94
Cameraman (salt–pepper)	95.03	99.09	98.45
vegetables (gaussian)	75.89	90.88	88.96
vegetables (salt–pepper)	94.99	98.15	97.50

Table 8.2 The noisy PSNR and the actual PSNR

Scrambled images	Noise types	PSNR (dB)
Image of lena	Salt and pepper	18.95
Image of lena	Gauss	20.78
Image of baboon	Salt and pepper	19.15
Image of baboon	Gauss	18.91
Image of cameraman	Salt and pepper	23.00
Image of cameraman	Gauss	23.86
Image of vegetables	Salt and pepper	18.97
Image of vegetables	Gauss	19.95

the suggested wavelet-based denosing technique for the sake of simplification in this quantum research. In our studies, we looked at both fixed (gauss) and random (salt and pepper) noise patterns. Importantly, we want to emphasize that this occurs without depending on the quantum noise, implying that the unpredictability about the outputs is an inherent aspect of this kind of quantum techniques. We'd like to repeat the measurement multiple times to be almost certain of the replies from quantum computers.

The fidelity metric values of different noisy images and the associated filtered images created by the suggested approach are shown in Table 8.1. According to Table 8.1, the QIFM grew for each image after filtering the gaussian and salt and pepper noises, respectively. In only two cases, however, QIFM achieved by the suggested approach is lesser than that obtained by the standard filtering scheme. However, in terms of the computational benefits of quantum parallelism, this minor variation may be insignificant. The results of the experiments show that the suggested method is effective (Tables 8.2 and 8.3). A detail comparison on computational complexity analysis with some benchmark methods are described in Table 8.4.

Table 8.3 The PSNR and MSE values after quantum wavelet transform

Image	Noise types	Tuning parameters	PSNR (dB)	Mean square error
Image of lena	Salt and pepper	rho = 0.05	37.12	0.0305
Image of lena	Gauss	mu = 0 and $\sigma^2 = 0.05$	36.07	0.0352
Image of baboon	Salt and pepper	rho = 0.05	29.11	0.0755
Image of baboon	Gauss	mu = 0 and $\sigma^2 = 0.05$	30.15	0.0769
Image of cameraman	Salt and pepper	rho = 0.05	35.01	0.0389
Image of cameraman	Gauss	mu = 0 and $\sigma^2 = 0.05$	33.25	0.0449
Image of vegetables	Salt and pepper	rho = 0.05	28.98	0.0811
Image of vegetables	Gauss	mu = 0 and $\sigma^2 = 0.05$	32.05	0.0548

Table 8.4 Computational time complexity analysis

$\text{ID}_{\text{non-local}}$ [2]	ID_{PCA} [3]	$\text{ID}_{\text{Curvelet}}$ [4]	ID-adaptive wavelet [5]	ID_{QPSO} [12]	$\text{Proposed}_{\text{tech}}$
$O(n^4)$	$O(n^3)$	$O(n^2 \log(n))$	$O(n^2)$	$O(n^3)$	$O(n^2)$

8.3 Quantum Image Edge Detection

Edge detection is a technique for enhancing the look of an area's or object's contours in the image so that they may be more easily identified. There have been numerous edge extraction techniques developed in both the classical [13–15] and quantum domains [16–18] over the past few decades. By computing the gradients of the image brightness of every pixel, they significantly influence the extraction of image edges. Furthermore, a $2^n \times 2^n$ resolution-based edge extraction, certain well-known classical edge extraction techniques demonstrate $O(2^{2n})$ computational complexity. Comparatively speaking to conventional techniques, the quantum image edge extraction technique makes use of the strong quantum properties to process every pixel at once, that can result in a notable speedup in the procedure of image edge extraction.

With the aid of the Sobel operator, the quantum image edge extraction technique QSobel was initially put forth by Zhang et al. [16]. QSobel uses the FRQI technique and superposition to calculate the Sobel gradients of all pixels at once. In the paper [16], a Sobel mask and NEQR combined method is proposed for extracting the edges of quantum images. With the use of an enhanced Kirsch operator, another quantum edge extraction method is introduced [17]. The quality metrics PSNR and MSE are used to assess the efficacy and efficiency of the suggested plan as well. The use of an artificial selection of threshold values is a limitation of all the approaches mentioned above. To address these issues, a method by combining the adaptive threshold value method and non-maximum suppression method is developed [18]. By establishing a certain gradient threshold, a neighborhood quantum feature object fetching method is established [19]. Feature point extraction is carried out using a straightforward neighborhood mask. By merging the zero-cross approach and the traditional Laplacian operator, a very new edge detection technique is suggested [20]. With the use of the parallelism of quantum processing, a new technique is constructed utilizing the Marr–Hildreth technique, which combines Gaussian–Laplacian filtering and zero-crossing extraction techniques [21]. The quantum adders and multipliers have been used to create a Gaussian–Laplacian filtering circuit. The quantum comparators have been used to build a zero-crossing circuit.

By combining yet another well-known classical Robinson operator with a revolutionary quantum image edge extraction methodology, we have addressed the problem with conventional classical image edge detection methods. This method is inspired by all of the edge extraction methods previously discussed. We predict that the proposed quantum operator will surpass other previous established operators. According to the Robinson edge detection operator's orientation, we can adjust the influence of image edge detection. In addition, the speedup, image quality, and edge detection accuracy are all improved using this technique. We construct and refine our suggested system after NEQR image representation and conduct a thorough study of the proposed quantum method. We discuss that our method outperforms the existing techniques with respect to the computational complexity, increased accuracy, and quality of resultant image at the conclusion of our result analysis section.

8.3.1 Background

We provide a brief study of the Robinson edge detection mask [8, 10, 22] and NEQR quantum image representation (QIR) models, which serve as the cornerstones of the suggested methodology in this part of the chapter.

8.3.1.1 NEQR Model

A novel enhanced quantum representation (NEQR) approach employs the base state of a sequence of qubits to save the grayscale value of every original pixel [4, 23]. In order to represent the color information and positioning information of the pixel, respectively, two entangled qubit sequences are used in NEQR. Consider the case where a grayscale image of dimension $2^n \times 2^n$ with a color range of 0 to 2^q can be represented by Eq. (8.1).

$$f(y, x) = C_{yx}^0, C_{yx}^1, \ldots, C_{yx}^{q-1}, \tag{8.19}$$

where, $C_{yx}^k \in [0, 1]$, $f(y, x) \in [1, 2^q - 1]$.

It is also possible to write the expression as Eq. (8.2).

$$\begin{aligned} |P\rangle &= \frac{1}{2^n} \sum_{y=0}^{2^n-1} \sum_{x=0}^{2^n-1} |f(y, x)\rangle |yx\rangle, \\ &= \frac{1}{2^n} \sum_{y=0}^{2^n-1} \sum_{x=0}^{2^n-1} \otimes_{i=0}^{q-1} |C_{yx}^i\rangle |yx\rangle \end{aligned} \tag{8.20}$$

As a result, the NEQR quantum image preparation requires substantially less time than the FRQI $O(qn \cdot 2^{2n})$ [23]. Equation (8.3) may be used to numerically represent a 2×2 NEQR image.

$$\begin{aligned} |P\rangle &= \frac{1}{2} |0\rangle |00\rangle + |100\rangle |01\rangle + |200\rangle |10\rangle + |255\rangle |11\rangle \\ &= \frac{1}{2} |00000000\rangle |00\rangle + |01100100\rangle |01\rangle + |11001000\rangle |10\rangle + |11111111\rangle |11\rangle \end{aligned} \tag{8.21}$$

8.3.1.2 Robinson Mask Classical Representation

One mask is rotated in this Robinson operator in each of the eight primary compass directions, including north, north-west, west, south-west, south, south-east, east, and north-east. To generate all of the direction masks in our study, we took one beginning mask (described below) and rotated it based on the 0th column. Although the Robinson operator demands more calculation than the Roberts and Sobel operators, it improves the gradient's precision. In Fig. 8.5, each Robinson operator template provides the edges in a particular direction. The integration of unitary quantum gates on the sequence of position qubits can be used to create the X-shift and Y-shift geometric transformations after describing a quantum image using the NEQR model. Due to the superposition of states, the position transformations of every pixel in the image are computed concurrently.

8.3.1.3 Location Transformations

Each pixel of the image can be moved in accordance with the X-shift and Y-shift location transformations, using the cyclic shift transformation given by Le et al. [5]. Equation (8.4) represents the details of position transformations,

-1	0	1
-2	0	2
-1	0	1

(a) North drection mask

0	1	2
-1	0	1
-2	-1	0

(b) North-West drection mask

1	2	1
0	0	0
-1	-2	-1

(c) West drection mask

2	1	0
1	0	-1
0	-1	-2

(d) South-West drection mask

1	0	-1
2	0	-2
1	0	-1

(e) South drection mask

0	-1	-2
1	0	-1
2	1	0

(f) South-East drection mask

-1	-2	-1
0	0	0
1	2	1

(g) East drection mask

-2	-1	0
-1	0	1
0	1	2

(h) North-East drection mask

Fig. 8.5 Classical Robinson mask with various directions

$$\begin{aligned}\mathrm{Sh}(x\pm)|P\rangle &= \frac{1}{2^n}\sum_{y=0}^{2^n-1}\sum_{x=0}^{2^n-1}\otimes_{i=0}^{q-1}|C_{yx}\rangle|y\rangle|(x\pm 1) \mod 2^n\rangle \\ \mathrm{Sh}(y\pm)|P\rangle &= \frac{1}{2^n}\sum_{y=0}^{2^n-1}\sum_{x=0}^{2^n-1}\otimes_{i=0}^{q-1}|C_{yx}\rangle|(y\pm 1) \mod 2^n\rangle|x\rangle\end{aligned} \quad (8.22)$$

Figure 8.6 depicts the quantum circuit for n-qubit shifting operation [24].

8.3.1.4 Quantum Parallel Addition Operation (QPADD)

In the paper [12], the quantum parallel adder included a detail of quantum cost and delay analysis using the Peres Gate is discussed. By combining of several image

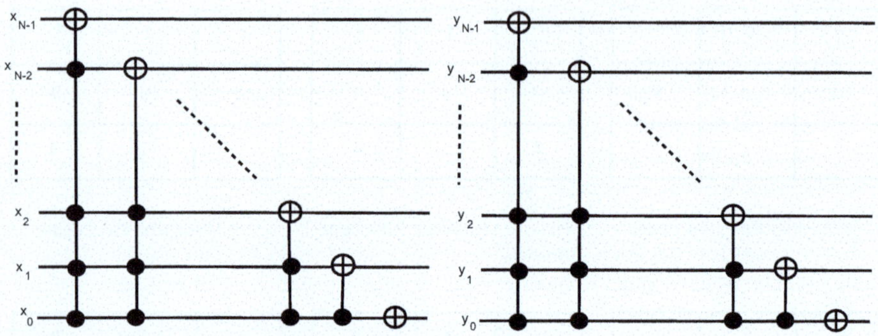

Fig. 8.6 Quantum circuits for shifting operation

pixels superposition value, the gradient of the image may be computed [25]. The resulting image state thus appears as follows: $|P_C\rangle = \text{QPADD}(|P_A\rangle, |P_B\rangle)$, where $|P_A\rangle, |P_B\rangle$ are subset of quantum images. Equation (8.5) represents as,

$$\begin{aligned} |P_{yx}\rangle &= \frac{1}{2^n}\sum_{y=0}^{2^n-1}\sum_{x=0}^{2^n-1}|C_{yx}\rangle|yx\rangle \\ &= \frac{1}{2^n}\sum_{y=0}^{2^n-1}\sum_{x=0}^{2^n-1}|A_{yx}+B_{yx}\rangle|yx\rangle \end{aligned} \tag{8.23}$$

8.3.1.5 Quantum Subtraction

A quantum block of subtractor circuit is designed in [13]. Furthermore, the operations of quantum addition and complement can be used to do subtraction operations in order to maintain simplicity [25]. So, the resultant image state looks like, $|P_C\rangle = \text{QPSUB}(|P_A\rangle, |P_B\rangle)$, where $|P_A\rangle, |P_B\rangle$ are subset of quantum images. Equation (8.6) represents as,

$$\begin{aligned} |P_{yx}\rangle = \text{QPADD}(|P_A\rangle, |P'_B\rangle) &= \frac{1}{2^n}\sum_{y=0}^{2^n-1}\sum_{x=0}^{2^n-1}|C'_{yx}\rangle|yx\rangle \\ &= \frac{1}{2^n}\sum_{y=0}^{2^n-1}\sum_{x=0}^{2^n-1}|A_{yx}+B'_{yx}\rangle|yx\rangle \end{aligned} \tag{8.24}$$

8.3.1.6 Double Operation (DO)

In this operation, we need to double the value of each pixel in order to calculate the gradient of the image using the Robinson mask. Assume that binary bits $|C\rangle = |c_{q-1}c_{q-2}\cdots c_1c_0 0\rangle$ and that we will perform the doubling operation by shifting 1-bit toward left direction.

$$\begin{aligned} |2\cdot C\rangle &= |c_{q-1}c_{q-2}\cdots c_1c_0 0\rangle \\ |0\rangle \otimes |C\rangle &= |0c_{q-1}c_{q-2}\cdots c_1c_0\rangle \end{aligned} \tag{8.25}$$

An additional qubit $|0\rangle$ is entangled in Eq. (8.25). The paper [22, 24, 26] discusses this DO operation in detail.

8.3.1.7 Thresholding Operation (U_{TH})

We need a quantum black box U_{TH} to do quantum classification. Here, "TH" represents threshold. However, if the entire n-length qubit sequences that are shorter than "TH" correspond to the first group in this thresholding process and the remaining to the second group. A starting auxiliary qubit $|0\rangle$ is used to hold the categorization output. $|c_{q+2}c_{q+1}\cdots c_1c_0\rangle|0\rangle$ is how it is written. The following section contains a detailed discussion of U_{TH}.

The proposed operator also includes 8 different types of 3×3 masks, though. The $|Ri\rangle$ is derived from the highest intensity among all permutations ($|R_0\rangle$, $|R_1\rangle$, $|R_2\rangle$, $|R_3\rangle$, $|R_4\rangle$, $|R_5\rangle$, $|R_6\rangle$, $|R_7\rangle$), such as by applying the X-shift and Y-shift transformations progressively to the experimented image $|P_{yx}\rangle$. Thus, the calculation of $|R_0\rangle$ is defined in Eq. (8.26),

$$R_0 = \begin{bmatrix} -1 & 0 & 1 \\ -2 & 0 & 2 \\ -1 & 0 & 1 \end{bmatrix} * \begin{bmatrix} P_1 & P_0 & P_7 \\ P_2 & P & P_6 \\ P_3 & P_4 & P_5 \end{bmatrix} \tag{8.26}$$

where * represents convolution operation. In the similar way, the remaining values of $|R_i\rangle$ can be computed. Thus, the $|R\rangle$ is evaluated through Eq. (8.27),

$$\begin{aligned} R_i &= \mathrm{MAX}(|R_0\rangle, |R_1\rangle, |R_2\rangle, \ldots, |R_7\rangle) \\ &= -P_1 - 2P_2 - P_3 + P_7 + 2P_6 + P_5 \\ &= (P_7 + 2P_6 + P_5) - (P_1 + 2P_2 + P_3) \end{aligned} \tag{8.27}$$

if $R \geq \mathrm{TH}$, the pixel is going to be regarded as a piece of the edge.
otherwise $R < \mathrm{TH}$, it won't be regarded as a component of the edge.

Therefore, using the aforementioned quantum image shift transforms operation, we can determine the gradient value of each pixel for 8 distinct directions. Equation contains all of the gradient calculations as well as the Eight directed Robinson operator templates and shifting operations (8.30).

8.3.1.8 Quantum Robinson Technique

S-1

Equation (8.28) produces 8 quantum images according to x-y shifting coordinate transformations.

$$
\begin{aligned}
&\text{Input} : |P_{yx}\rangle = \frac{1}{2^n}\sum_{y=0}^{2^n-1}\sum_{x=0}^{2^n-1}|C_{yx}\rangle|yx\rangle,\\
&\text{I} : |P_{y-1x}\rangle = \frac{1}{2^n}\sum_{y=0}^{2^n-1}\sum_{x=0}^{2^n-1}|C_{y-1x}\rangle|yx\rangle,\\
&\text{II} : |P_{y-1x-1}\rangle = \frac{1}{2^n}\sum_{y=0}^{2^n-1}\sum_{x=0}^{2^n-1}|C_{y-1x-1}\rangle|yx\rangle,\\
&\text{III} : |P_{yx-1}\rangle = \frac{1}{2^n}\sum_{y=0}^{2^n-1}\sum_{x=0}^{2^n-1}|C_{yx-1}\rangle|yx\rangle,\\
&\text{IV} : |P_{y+1x-1}\rangle = \frac{1}{2^n}\sum_{y=0}^{2^n-1}\sum_{x=0}^{2^n-1}|C_{y+1x-1}\rangle|yx\rangle,\\
&\text{V} : |P_{y+1x}\rangle = \frac{1}{2^n}\sum_{y=0}^{2^n-1}\sum_{x=0}^{2^n-1}|C_{y+1x}\rangle|yx\rangle,\\
&\text{VI} : |P_{y+1x+1}\rangle = \frac{1}{2^n}\sum_{y=0}^{2^n-1}\sum_{x=0}^{2^n-1}|C_{y+1x+1}\rangle|yx\rangle,\\
&\text{VII} : |P_{yx+1}\rangle = \frac{1}{2^n}\sum_{y=0}^{2^n-1}\sum_{x=0}^{2^n-1}|C_{yx+1}\rangle|yx\rangle,\\
&\text{VIII} : |P_{y-1x+1}\rangle = \frac{1}{2^n}\sum_{y=0}^{2^n-1}\sum_{x=0}^{2^n-1}|C_{y-1x+1}\rangle|yx\rangle,\\
&\text{IX} : |P_{yx}\rangle = \frac{1}{2^n}\sum_{y=0}^{2^n-1}\sum_{x=0}^{2^n-1}|C_{yx}\rangle|yx\rangle,
\end{aligned}
\tag{8.28}
$$

$$
\begin{aligned}
\text{Imageset} = [&|P_{x-1,y-1}\rangle, |P_{x,y-1}\rangle, |P_{x+1,y-1}\rangle, |P_{x-1,y}\rangle, |P_{x,y}\rangle, |P_{x+1,y}\rangle,\\
&|P_{x-1,y+1}\rangle, |P_{x,y+1}\rangle, |P_{x+1,y+1}\rangle]
\end{aligned}
\tag{8.29}
$$

S-2

Equation (8.32) determines the pixel's gradient. Therefore, we are simply displaying the $|R_0\rangle$ gradient computation here.

$$
\begin{aligned}
|M_0\rangle &= \text{QPADD}(|P_{x+1,y-1}\rangle, 2|P_{x+1,y}\rangle) \\
|M_1\rangle &= \text{QPADD}(|M_0\rangle, |P_{x+1,y+1}\rangle) \\
|N_0\rangle &= \text{QPADD}(|P_{x-1,y-1}\rangle, 2|P_{x-1,y}\rangle) \\
|N_1\rangle &= \text{QPADD}(|N_0\rangle, |P_{x-1,y+1}\rangle) \\
|R_0\rangle &= \text{QPSUB}(|M_1\rangle, |N_1\rangle)
\end{aligned}
\tag{8.30}
$$

Calculate R_{MAX} using a template operation as an analogy. Equation (8.31) represents the operated image.

$$
|P'_{x,y}\rangle = \frac{1}{2^n}\sum_{y=0}^{2^n-1}\sum_{x=0}^{2^n-1}|C_{R_{\text{MAX}}}(yx)\rangle|yx\rangle \tag{8.31}
$$

S-3

Consequently, the computed intensity gradient $|R_{yx}\rangle \in [0, \ldots, 2^{(q+n)-1}]$ can be obtained by setting a threshold in the threshold operation U_{TH} of the original image can be categorized (8.5).

$$
U_{\text{TH}}\left(\sum_{yx=0}^{2^{2n}-1}|R_{yx}\rangle|0\rangle\right) = \sum_{Ryx\geq \text{Th}}|R_{yx}\rangle|1\rangle + \sum_{Riyx<\text{Th}}|R_{iyx}\rangle|0\rangle \tag{8.32}
$$

S-4

At last, the resultant image is,

$$
|\text{Th}\rangle = \frac{1}{2^n}\sum_{i=0}^{2^{2n}-1}|\text{Th}_{yx}\rangle|yx\rangle,\ \text{Th}_{yx} \in 0, 1 \tag{8.33}
$$

Equation (8.33) indicates that $\text{Th}_{yx} = 1$ is an edge point.
if $|R\rangle \geq |\text{Th}\rangle$, then $|C_i\rangle = |0\rangle$
else, $|C_i\rangle = |1\rangle$.

8.3.2 Analysis of Results

Table 8.5 compares the computational complexity of our suggested method with a few other widely used classical methods.

Figure 8.7 shows an overall quantum block circuit of the proposed method. The three experimented medical images are X-rays of the human body's skeleton, MR images of the knee, and ankle joint, respectively. In order to recognize the hard features like bones and separating those features from the background depicted in

Table 8.5 Complexity comparison

Methods	Representation model	Model complexity	Method complexity
Zhang [27]	NEQR	$O(qn2^{2n})$	$O(n^2 + q^2)$
TSeng [7]	Qubit lattice	$O(2^{2n})$	$O(2^{2n})$
Fu [28]	Qubit lattice	$O(2^{2n})$	$O(2^{2n})$
QSobel-Zhang [29]	FRQI	$O(2^{4n})$	$O(n^2)$
Sobel-Fan [24]	NEQR	$O(qn2^{2n})$	$O(n^2 + 2^{q+4})$
Zhou-Prewitt [26]	NEQR	$O(qn2^{2n})$	$O(n^2 + 2^{q+3})$
Canny detector [8]	–	–	$O(2^{2n})$
Xu-Kirsch [25]	NEQR	$O(qn2^{2n})$	$O(n^2 + 2^{q+3})$
Proposed scheme	NEQR	$O(qn2^{2n})$	$O(n^2 + 2^{q+3})$

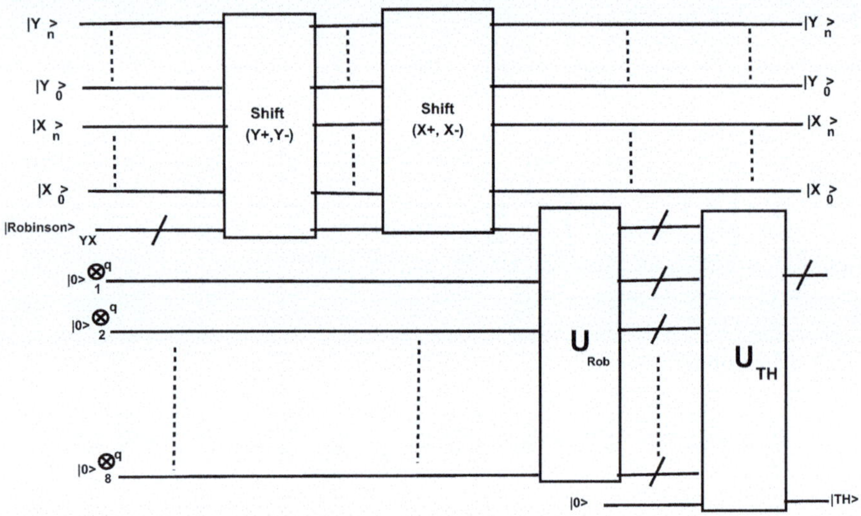

Fig. 8.7 Proposed quantum block circuit

Table 8.6 PSNR, MSE and execution time comparison

Methods	MSE	PSNR (dB)	Running time (s)
Typical Sobel [24]	4.61	35.71	0.919
Typical Prewitt [26]	4.79	37.99	0.90
Typical Canny [8]	4.89	39.7	0.929
Quantum Shannon [16]	4.55	35.39	0.851
QGA + SVM [17]	4.51	35.91	0.88
Proposed method	4.55	35.89	0.82

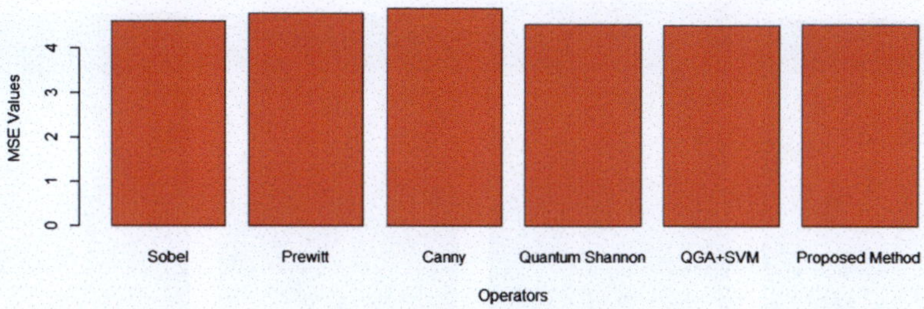

(a) Mean MSE Values

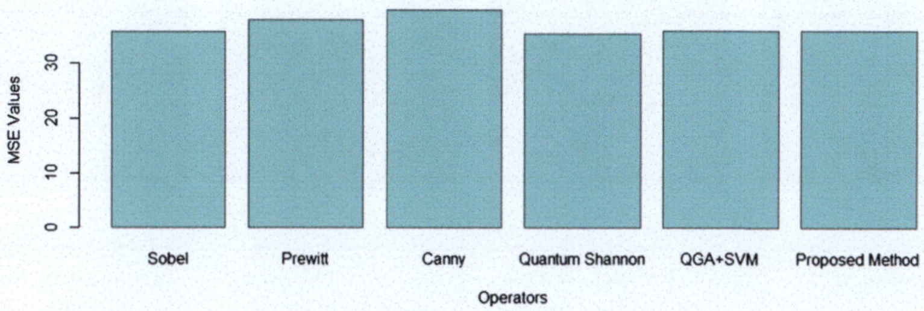

(b) Mean PSNR Values

Fig. 8.8 Graphical comparison on mean PSNR and MSE values

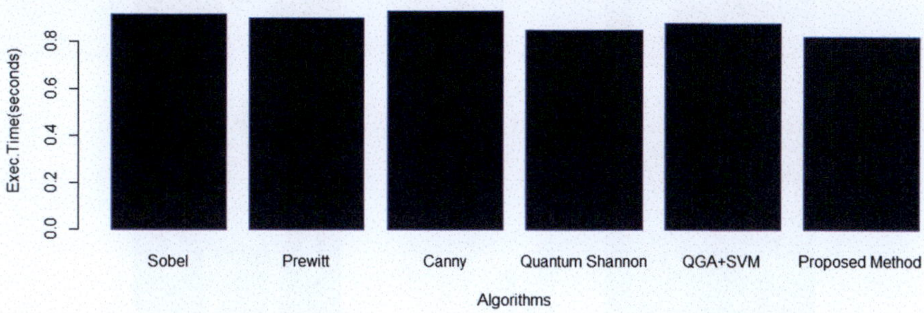

Fig. 8.9 Comparison on execution time

Figs. 8.10, 8.11, and 8.12, the suggested method outperforms existing well-known algorithms. Based on the experimented images, it is evident that the proposed method outperforms state-of-the-art classical techniques in terms of edge intensity and identifying real from falsely edges. Canny exhibits an excessive sensitivity to softening areas, while Prewitt and Sobel have a weakness in edge brightness, which is fixed in the proposed method. The average MSE and PSNR values in Table 8.6 and Fig. 8.8 demonstrate the quality of the resultant experimented images, respectively. Figure 8.9 represents the comparison on execution time. After using the suggested Robinson edge extraction approach, the final image's PSNR is 35.91 dB, indicating fine quality

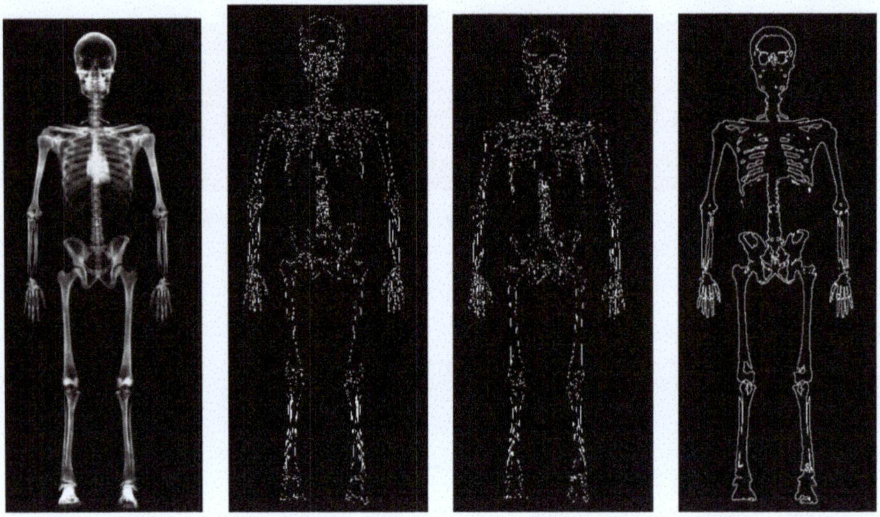

(a) Original Skeleton gray image

(b) Skeleton image of Prewitt operation with threshold=0.128

(c) Skeleton image of Sobel operation with threshold=0.128

(d) Skeleton image of quantum Shannon operation with threshold=0.128

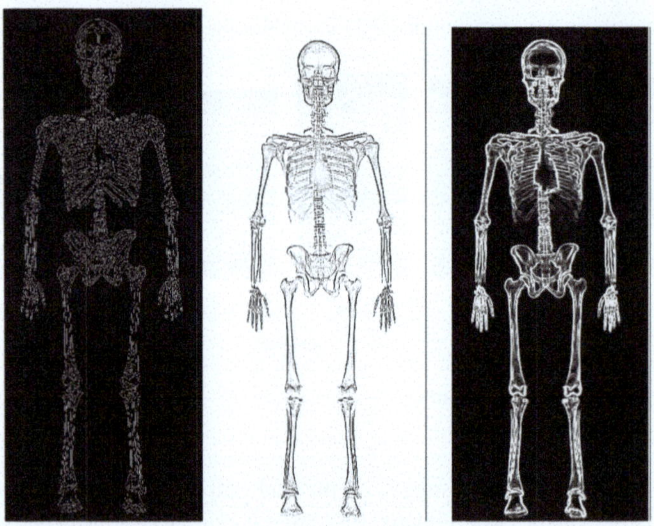

(e) Skeleton image of classical Canny operation with threshold=0.128

(f) Skeleton image of QGA with SVM operation

(g) Skeleton image of proposed quantum Robinson operation with threshold=0.128

Fig. 8.10 Simulation outcomes on the experimented skeleton image

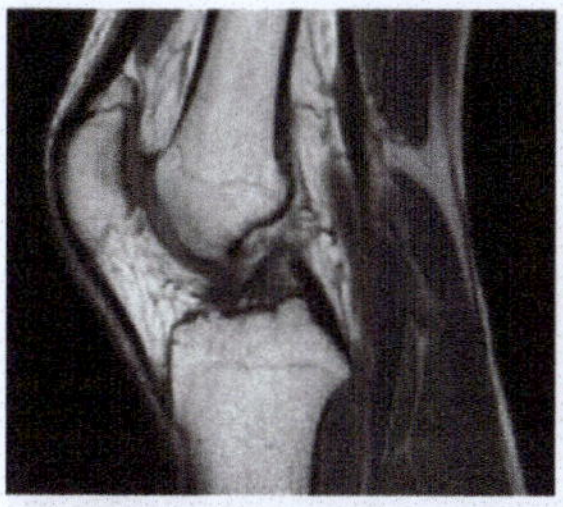

(a) Original Knee gray image

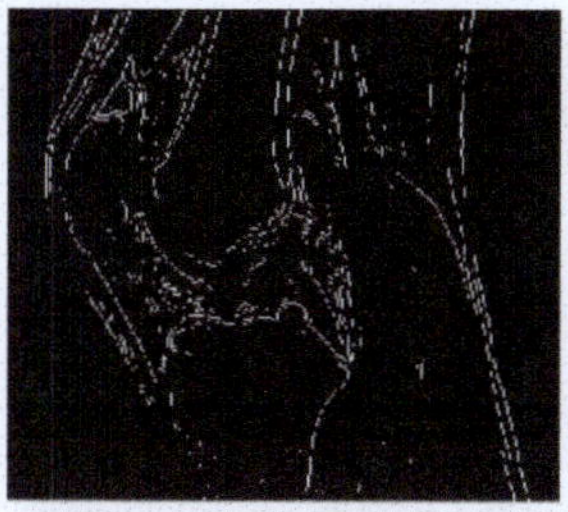

(b) Knee image of Prewitt operation with threshold=0.120

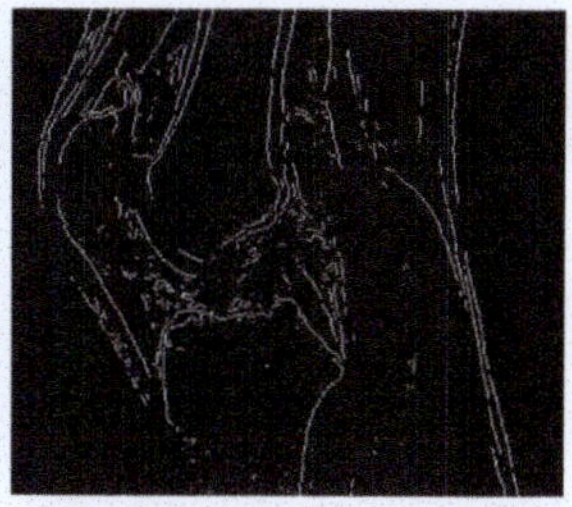

(c) Knee image of Sobel operation with threshold=0.120

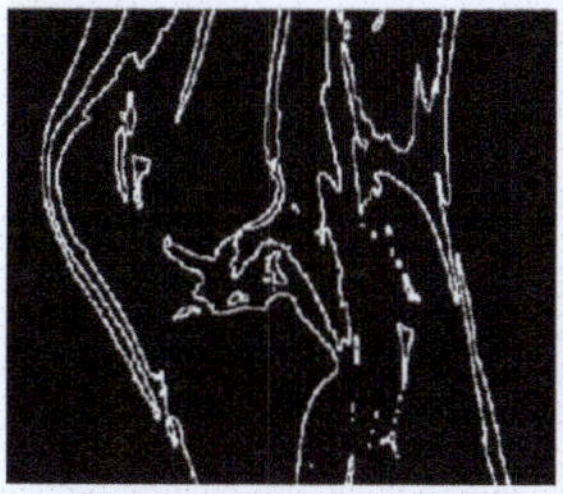

(d) Knee image of quantum Shannon operation with threshold=0.120

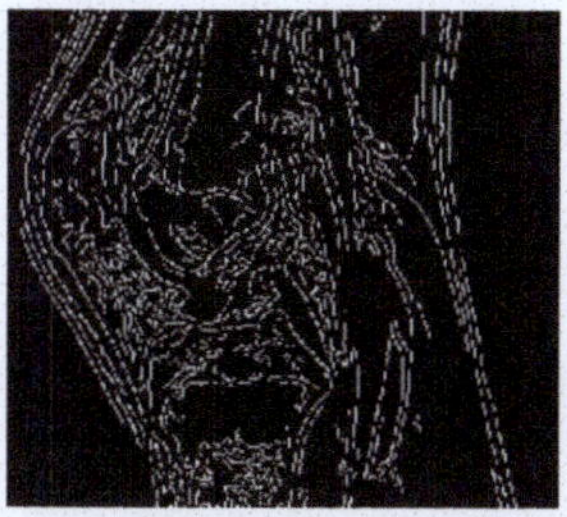

(e) Knee image of classical Canny operation with threshold=0.120

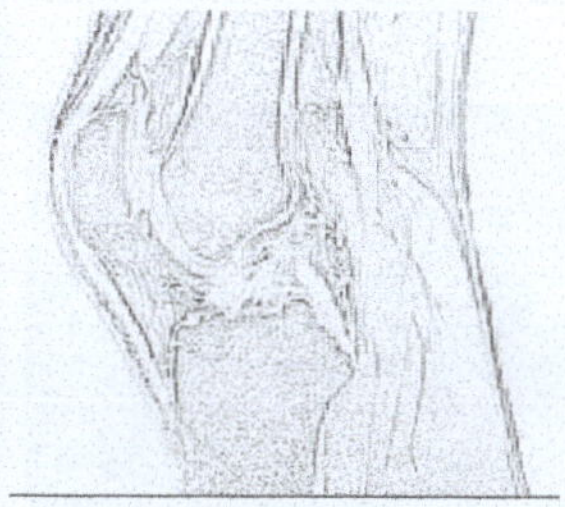

(f) Knee image of QGA with SVM operation

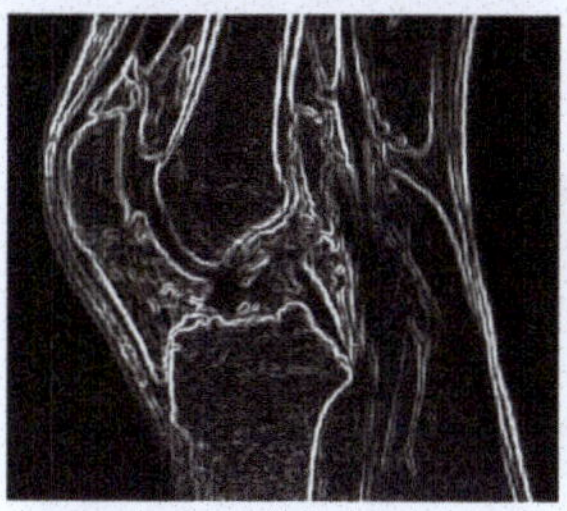

(g) Knee image of proposed quantum Robinson operation with threshold=0.120

Fig. 8.11 Simulation outcomes on the experimented knee image

of image. But, the traditional Sobel and quantum Shannon algorithms outperform the previously mentioned approach in terms of performance. However, when compared to other methods, our solution provides the top speedup.

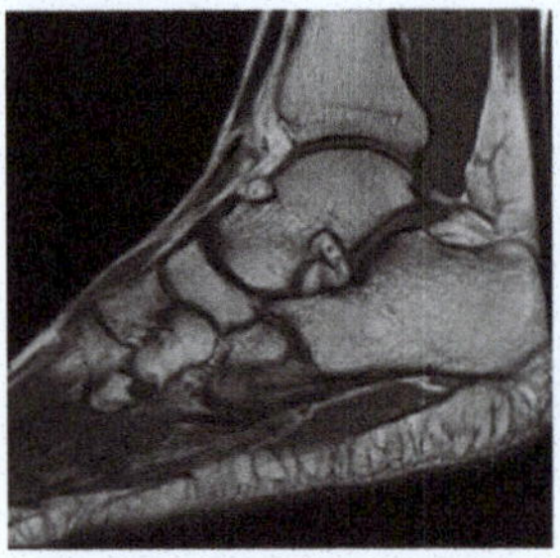

(a) Original ankel joint gray image

(b) Ankel joint image of Prewitt operation with threshold=0.135

(c) Ankel joint image of Sobel operation with threshold=0.135

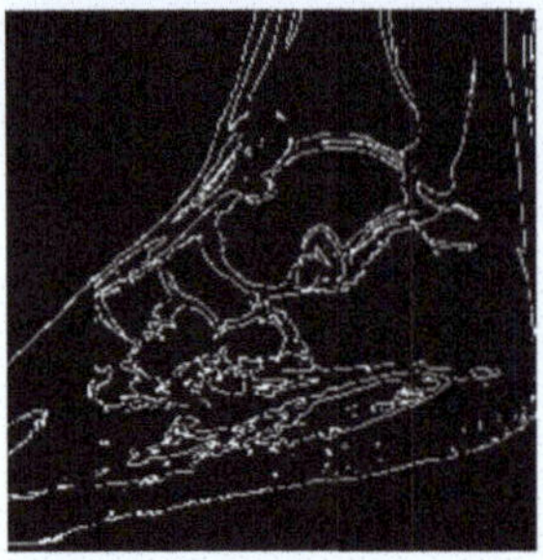

(d) Ankel joint image of quantum Shannon operation with threshold=0.135

(e) Ankel joint image of classical Canny operation with threshold=0.135

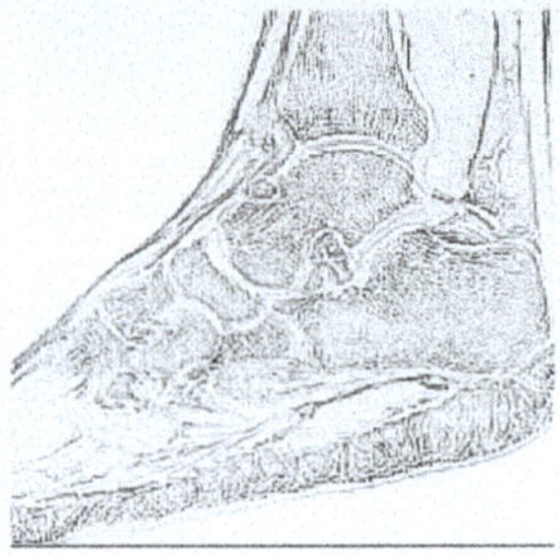

(f) Ankel joint image of QGA with SVM operation

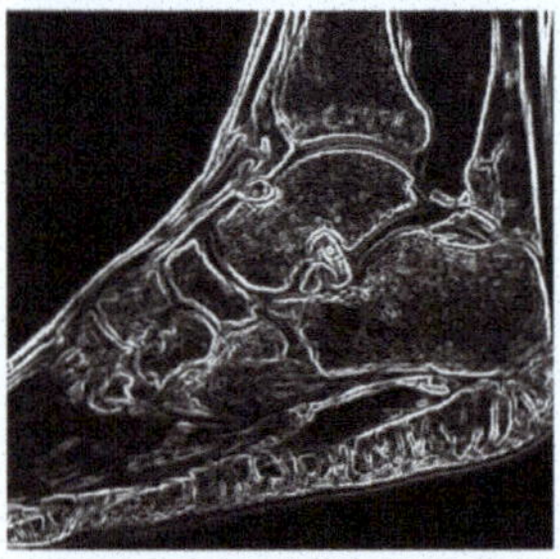

(g) Ankel joint image of proposed quantum Robinson operation with threshold=0.135

Fig. 8.12 Simulation outcomes on the experimented ankle joint image

8.4 Conclusion

In this chapter, the two quantum image handling techniques are described that perform two significant operations (denoising and edge detection) on images. It gives a different essence of how quantum principles can be used for doing better image processing tasks than some well-known classical and quantum counterparts. It shows the major comparisons in terms of PSNR, MSE, and speedup factors. These two works represent the doorsteps of quantum-classical hybrid techniques suitable in the NISQ era.

References

1. Choudhury A, Samanta S, Dey N, Ashour AS, Bălas-Timar D, Gospodinov M, Gospodinova E (2015) Microscopic image segmentation using quantum inspired evolutionary algorithm. J Adv Microsc Res 10(3):164–173
2. Wang J, Guo Y, Ying Y, Liu Y, Peng Q (2006) Fast non-local algorithm for image denoising. In: 2006 international conference on image processing. IEEE, pp 1429–1432
3. Chen G, Qian SE (2011) Denoising of hyperspectral imagery using principal component analysis and wavelet shrinkage. IEEE Trans Geosci Remote Sens 49(3):973–980
4. Starck JL, Candès EJ, Donoho DL (2002) The curvelet transform for image denoising. IEEE Trans Image Process 11(6):670–684
5. Cai Z, Cheng TH, Lu C, Subramanian KR (2001) Efficient wavelet based image denoising algorithm. IEEE Electron Lett 37(11):670–684
6. Manta VI, Caraiman S (2013) Quantum image filtering in the frequency domain. Adv Electr Comput Eng 13(3):77–84
7. Yuan S, Mao X, Zhou J, Wang X (2017) Quantum image filtering in the spatial domain. Int J Theor Phys 56(8):2495–2511
8. Li P, Liu X, Xiao H (2018) Quantum image median filtering in the spatial domain. Quantum Inf Process 17(3):49
9. Fijany A, Williams CP (1998) Quantum wavelet transforms: fast algorithms and complete circuits. In: NASA international conference on quantum computing and quantum communications. Springer, Berlin, Heidelberg, pp 10–33
10. Le PQ, Dong F, Hirota K (2011) A flexible representation of quantum images for polynomial preparation, image compression, and processing operations. Quantum Inf Process 10(1):63–84
11. Vedral V, Barenco A, Ekert A (1996) Quantum networks for elementary arithmetic operations. Phys Rev A 54(1):147–153
12. Djemame S, Batouche M, Oulhadj H, Siarry P (2018) Solving reverse emergence with quantum PSO application to image processing. Soft Comput 1–15
13. Robinson GS (1977) Edge detection by compass gradient masks. Comput Graph Image Process 6(5):492–501
14. Simangsong PBN (2020) Image edge detection using Robinson operator. INFOKUM 8(2):31–36
15. Malik S, Kumar T (2016) Comparative analysis of edge detection between gray scale and color image. Commun Appl Electron (CAE). ISSN: 2394-4714
16. Fan P, Zhou RG, Hu W, Jing N (2019) Quantum image edge extraction based on classical Sobel operator for NEQR. Quantum Inf Process 18(1):24
17. Xu P, He Z, Qiu T, Ma H (2020) Quantum image processing algorithm using edge extraction based on Kirsch operator. Opt Express 28(9):12508–12517

18. Zhou RG, Yu H, Cheng Y, Li FX (2019) Quantum image edge extraction based on improved Prewitt operator. Quantum Inf Process 18(9):261
19. Zhang Y, Lu K, Xu K, Gao Y, Wilson R (2015) Local feature point extraction for quantum images. Quantum Inf Process 14(5):1573–1588
20. Fan P, Zhou RG, Hu WW, Jing N (2019) Quantum image edge extraction based on Laplacian operator and zero-cross method. Quantum Inf Process 18(1):27
21. Li P, Shi T, Lu A, Wang B (2020) Quantum implementation of classical Marr-Hildreth edge detection. Quantum Inf Process 19(2):1–26
22. Zhang Y, Ni Q (2020) Recent advances in quantum machine learning. Quantum Eng 2(1):e34
23. Singh L, Janghel R (2019) Image denoising techniques: a brief survey. Harmony search and nature inspired optimization algorithms. Springer, Singapore, pp 731–740
24. Sleeman J, Dorband J, Halem M (2020) A hybrid quantum enabled RBM advantage: convolutional autoencoders for quantum image compression and generative learning. In: Quantum information science, sensing, and computation XII, vol 11391. International Society for Optics and Photonics, p 113910B
25. Henderson M, Shakya S, Pradhan S, Cook T (2020) Quanvolutional neural networks: powering image recognition with quantum circuits. Quantum Mach Intell 2(1):1–9
26. Kak S (2000) Active agents, intelligence and quantum computing. Inf Sci 128:1–17
27. Caraiman S, Manta VI (2013) Image representation and processing using ternary quantum computing. In: 11th international conference on adaptive and natural computing algorithms, Lausanne, Switzerland, pp 366–375
28. Yan F, Chen K, Iliyasu AM, Hirota K (2020) Circuit-based modular implementation of quantum ghost imaging. IEEE Access 8:23054–23068
29. Ciliberto C, Herbster M, Ialongo AD, Pontil M, Rocchetto A, Severini S, Wossnig L (2018) Quantum machine learning: a classical perspective. Proc Roy Soc A: Math Phys Eng Sci 474(2209):20170551

Part IV
Computations for Various Data Applications and Future Work

Chapter 9
Challenges and Future Research Directions on Data Computation

9.1 Introduction

In this chapter, we have discussed the various applications, current challenges, and expected future research directions in data computation with respect to the big data, IoT, cloud computing, quantum computing, and biological computing frameworks. The relationship among these frameworks and their impact on each other are also described in the below sections.

9.2 Challenges and Future Research Directions for Big Data Analysis

Big data is enormous, intricate datasets. The intricacy of the data generated is further increased by the fact that it is highly dynamic. It is necessary to process the raw data in order to derive value from it. This results in difficulties processing massive data and related business problems. The amount of data produced globally is increasing dramatically. Almost all industries, including healthcare, automotive, finance, and transportation, rely on this information to improve their operations and business plans [1, 2]. Airlines, for instance, conduct millions of transactions per day and have set up data warehouses to store data in order to benefit from machine learning techniques and obtain statistical insight that would aid in business initiatives. To boost efficiency, the public administration sector also makes use of information patterns drawn from data generated by population segments with varying ages. Additionally, many scientific disciplines are now data-driven and explore the knowledge gained from these data. The typical method for managing and analyzing massive data has been cloud computing. Even with all the benefits of combining big data and cloud computing, there are still a number of difficulties with data transfer, storage, transformation, quality, privacy, and governance [3, 4].

S. Chakraborty and L. Dey, *Computing for Data Analysis: Theory and Practices*,
Data-Intensive Research, https://doi.org/10.1007/978-981-19-8004-6_9

Data Transmission

Datasets are expanding rapidly. The rate at which these real-time data is sent via communication networks has expanded along with their size. While some scientific applications may have to transmit gigabytes of data files that may take longer to traverse the network, the transmission of health information between healthcare professionals typically involves high-definition medical photographs that are sent electronically. In streaming applications, the transmission speed is just as important as the actual data packets' order in the stream. Cloud data stores are used to store data, but there are issues with network bandwidth, latency, throughput, and security [5].

Data Acquisition and Storage

Before data can be kept in any data warehouses or storage systems, it must first be collected from various sources, filtered, and cleaned. The sheer amount, increased velocity, and variety of the big data are the major traits that make acquisition difficult. This necessitates more flexible data collection, filtering, and cleaning algorithms that guarantee faster data acquisition. Once obtained, data must be kept in large-capacity databases that permit dependable access to the data. There are two types of storage technologies available right now: Direct attached storage (DAS) and network attached storage (NAS).

Data Curation

It describes the active and continuing management of data throughout every stage of its lifecycle, from creation or ingestion through archiving or being no longer relevant and being removed. To ensure that data is safely saved and retrievable, this procedure involves several stages of transformation. To curate data, organizations need to invest in the right personnel and provide them the appropriate resources. Improvements in high-quality data quantification will result from such an investment in data curation.

Scalability

Scalability is the capacity to offer resources in a way that is acceptable for meeting business needs. It is a predetermined capacity that may increase as necessary. It is static and primarily manual. To accommodate data changes, most big data systems need to be elastic. There is both vertical and horizontal scalability at the platform level. It becomes difficult to exponentially extend the cloud's capacity in order to supply storage and handle too many people who are connected to it at once when the number of cloud users and data grows quickly.

Elasticity

The ability of the cloud to maintain optimal performance regardless of computing workloads while lowering operational costs is referred to. Elasticity uses replication, migration, and scaling strategies in real-time to adapt to changes in data load without interrupting service. Instead of being automated, the majority of these are manual.

Availability

The term "availability" describes the systems' on-demand accessibility to authorized users. Allowing users to quickly access one or more data services is one of the main features of cloud service providers. More real-time system availability would be sought after as business models changed.

Data Integrity

In order to prevent misuse, data integrity refers to changes made to the data only by authorized users. Users of cloud-based applications can store and control their data in cloud data centers, but these applications must protect the integrity of the data. Since consumers might not have direct access to the data, the cloud ought to offer tools for verifying its accuracy.

Security and Privacy

It is crucial to maintain the security of the data kept in the cloud. Personal and sensitive data stored in the cloud ought to be designated as being intended solely for internal use and not to be disclosed to outside parties. This would be a significant issue when offering tailored and location-based services since access to personal data is necessary to get useful results. Every action, including data transmission over the network, system interconnection over the network, and mapping virtual machines to their corresponding physical counterparts, must be conducted in a secure manner.

Heterogeneity

Big data is enormous and varied. Data in various formats, including semi-structured and unstructured data from multiple sources, must be handled by cloud computing systems. Unstructured data such as documents, images, audio, and video can be challenging to search for and analyze. It can be very challenging to aggregate and reconcile all the unstructured data in real time so that it can be used to generate reports.

Data Governance and Compliance

Data governance outlines the exercise of control and responsibility over the handling of data that is required as well as individual accountability for attaining corporate goals. Data policies must be established regarding the type of data that must be stored and the various constraint models that restrict access to the underlying data. It is difficult to define stable data policies in the face of growing data volume and need for quicker and better data management technology, and such policies may have unintended consequences.

Data Uploading

It refers to how simple it is to upload huge data collections to the cloud. Typically, data is uploaded via the Internet. Data upload speed is also influenced by network bandwidth and security. In order to reduce upload times and offer a secure method of transferring data onto the cloud, this demands for advancement and effective data uploading algorithms.

Data Recovery

In situations like data loss due to corruption or virus assault, it refers to the processes and methods by which the data can be restored to its original state. It is vital to select a subset of data that is valuable to the company for backup because performing regular backups of petabytes of data is time-consuming and expensive. If this portion of the data is lost or corrupted, it will take weeks to reconstruct the missing data at these enormous scales, which will extend the users' downtime.

Data Visualization

Data visualization is a quick and simple technique to graphically depict complicated ideas for improved comprehension and intuition. It must find diverse relationships and patterns concealed by vast amounts of data. Traditional graphical representations can be used for organized data, but it can be challenging to display huge data that is very diverse, uncertain, and semi-structured or unstructured in real time. Massive parallelization is required to handle such massive dynamic data, which is difficult to visualize [6].

9.2.1 Challenges

Big data is a broad term for enormous amounts of data that are extremely challenging to analyze and store. Utilizing a variety of methodologies, big data analytics is used to examine big data and transform it into processed and usable form by deleting redundant and unimportant data. Big data processing is not something that can be done manually. Big data analytical techniques provide a number of benefits, but they also present a number of obstacles, which is a very severe problem. Some of the problems involve data security and privacy, intricate coding, challenging implementation, etc. Researchers should concentrate more on coming up with the best, most efficient ways to address those problems [1]. Three categories can be used to generally classify the difficulties.

- The key challenge is managing the storage, processing, and identification of hidden patterns in big data. The following issues with data need to be prioritized: storage of data; data velocity; different types of data; data computational power; data understandability; data quality; and efficient representation of data.
- Big data technologies are now present in many business sectors. Therefore, it is crucial for them to pick the best big data analytics methodologies and to profit the most from them. The business must handle the adjustments promptly. Some of the difficulties include: leadership quality; talent management skills; technology management; decision-making skills; and organizational climate.

- Although the specialists can fully examine the model based on prior experiences, there are still some areas where they fall short, which makes processing big data challenging. Choosing the best model for the data analysis is a difficult undertaking that requires careful consideration. The following procedural obstacles must be overcome: gathering data from multiple sources; converting obtained data into analyzable form; and identifying hidden patterns in the data, examining the range of methods necessary to filter the data.

9.2.2 *Future Research Directions*

- Today's most of the complex problems of big data can be efficiently solved with the help of the powerful quantum computing technology due to its exponentially larger memory capacity and parallel computations. Quantum computation is represented by quantum bits or qubits instead of classical bits. This means that bigger scale quantum computers can address many big data issues much more quickly than classical computers. However, we still have to wait because the real large quantum computer is not physically available in the market. Therefore, building a quantum computer and enabling quantum computing to address large data issues is a task for this generation [7].
- Cloud computing aids in creating a business model for all types of applications. The ability to store large amounts of data on the cloud is another benefit of cloud computing. The obvious one is the expense and time involved in uploading and downloading large amounts of data to and from the cloud. Otherwise, it becomes challenging to manage the hardware and computing distribution. The storage of data from real world and privacy challenges related to storing on public servers are the main problems, though. Big data combined with cloud computing will reach a top level of growth as a result of all these problems [5, 8, 9].
- Biological computing is closely connected with big data system. Biological computing is self-organized and responsible to store, retrieve, and process DNA, RNA, or protein kind of structures and it is more suitable for big data kind of applications. Techniques for computing that are bioinspired play a crucial part in intelligent data analysis and their use with large data. These algorithms' optimization applications make it possible to undertake data mining for huge datasets. When resolving issues with service provision, their simplicity and quick convergence to the best solution are their greatest advantages. We can infer from the conversations that bioinspired computing models offer wiser interactions, help in addressing ambiguities, and account for inevitable data losses. Therefore, it is anticipated that bioinspired computing may in the future significantly aid in handling massive data [10, 11].

9.3 Challenges and Future Research Directions for IoT Data Analysis

Even if the present IoT enabler technologies have advanced significantly in recent years, there are still many issues that need to be resolved, opening the door for new research facets. Numerous research difficulties are certain to emerge since the Internet of Things concept results from diverse technologies that are employed in data sensing, collection, action, processing, inference, transmission, notification, management, and storage. Numerous research organizations have been founded and are still being formed around the world with the primary goal of advancing IoT-related studies [12]. IoT processes, technologies, and items that can be connected continue to take on new dimensions as more and more research investigations are done, opening the door for even more IoT application features. We have enlisted some major challenges facing by the IoT researchers worldwide [13].

- IoT is a collection of numerous heterogeneous devices, networks, and communication protocols. That is not how any clouds can interact with heterogeneous (different) modules or communication protocols. It makes different types of gadgets more ambiguous (via different communication technologies). End-to-end resource lag and dishonest network behavior are the results.
- IoT devices and technologies are dependent on the basic foundation of wireless sensor networks (WSN). Therefore, the threats related to privacy and security issues are automatically fetched from the WSN into the architecture of IoT. Numerous assaults frequently take advantage of flaws in certain gadgets to access their systems, rendering safe equipment susceptible. This security gap further inspires complete security solutions, which include frameworks that help programmers create secure systems on heterogeneous devices, research that is effective in applied cryptography for data and system security, and non-cryptographic security measures. It is necessary to perform additional research on cryptographic security services that can function on IoT devices with limited resources. Due to the lack of acceptable user interfaces on practically all IoT devices, this would allow a variety of experienced individuals to safely operate and install IoT systems. Additional areas including communication secrecy, the reliability, authenticity, integrity, and supplemental protection criteria should be included in addition to the IoT's features. Furthermore, to improve the new joint notion between block chain and IoT technologies, privacy and security plays a vital role [14].
- The diverse nature of IoT and the vast amount of data generated, especially in this big data era, make the process for processing, analysis, and data handling extremely difficult [15]. The majority of systems currently use centralized systems to offload data and perform computationally demanding operations on a global cloud platform. However, there is a persistent worry that traditional cloud architectures would not be able to transfer the enormous sizes of data which are created and absorbed by IoT instruments, as well as support the additional computational load that goes along with it and simultaneously meet timing requirements. To

address this issue, the majority of systems rely on currently available technologies like fog computing and mobile cloud computing, both of which are dependent on edge processing. Applying information-centric networking (ICN) in the IoT is another research area in data management. These information-centric systems found to be highly beneficial. In addition to being vital to the success of IoT, data analysis, and its context present significant hurdles. To accomplish intelligent IoT functions, data must be used wisely once it has been acquired [14].

- Even while monitoring and sensing technologies have advanced greatly, they are nevertheless constantly changing, with an emphasis on usage of less energy and shape. Energy efficiency is essential since tags and sensors are sometimes required to operate continuously to capture real-time data, especially for extending lifetime. At the same time, new developments in nanotechnology and miniaturization have made it possible to create nanoscale actuators and sensors.
- Presently, a number of standards are used by different sectors to serve their applications. Due to the vast numbers and types of data as well as the heterogeneous devices, using common interface in such diverse entities is incredibly vital, and especially more so for applications that permit cross-organizational, in addition to a broad variety of system limits. As a result, the IoT systems are intended to support even greater levels of interoperability.
- IoT systems integration and testing will be difficult, to say the least, due to the different platforms, protocols, and APIs involved. The ambiguity surrounding changing standards will almost certainly hamper uptake. The quick creation of APIs would probably use up unplanned development resources, which will limit the ability of project teams to provide essential new functionality. Schedules will certainly slide and time to revenues will likely be slowed down by slower adoption and unanticipated development resource requirements, necessitating more investment for IoT initiatives and longer "runways" for companies.
- The Internet of Things can be useful to governments and other public institutions. Better remote care for the elderly in their homes, for instance, can lower health and long-term care expenses. Based on the data collected from thousands of drivers, road safety can be increased. Dimming lights on open highways can increase street lighting's effectiveness.
- Customers may receive more tailored product or service offers depending on their activities or location. When their linked car proposes an alternate route based on traffic observed by other vehicles, they can drive more efficiently by avoiding congestion. They can save money by using less energy or by having their auto insurance premiums reduced in response to their proven safe driving habits. Wearable technology that monitors the elderly in the home or provides input on health can make them healthier, safer, and more independent.
- Businesses can better serve customers by observing how they act, and they can also identify requirements for new goods or services. They can secure vehicles and equipment with location trackers and remote locking devices, protect buildings with remote security, and guarantee that delicate products (like medications) are regularly stored in the proper conditions. The use of smart meters by utilities to

reduce waste or loss or the provision of just-in-time preventative maintenance by equipment sellers are two examples of how they might become more effective.

- With intelligent irrigation that supplies water only when and where it is needed, farmers may be more productive. New business strategies that sell results rather than merely equipment could increase profits.
- Middleware used today in IoTs is essential for handling domain-specific requirements. It can process horizontal data flow between devices on many (multiple) platforms; therefore, development is required.
- IoT networks frequently have a staggeringly large number of nodes. Data related to them and the associated equipment must both be recoverable. In these circumstances, the shared identifier is necessary for the point-to-point network configuration to be successful (i.e., using IPv4 address mechanism, 4-byte address to each node). But because there are so many devices connected to the Internet today, there are not enough IPv4 addresses available to give each and every connected device an address. In order to fix this issue, we therefore need a new addressing mechanism or policy, and IPv6 provides a powerful tool. Be aware that most current systems employ the IPv4 addressing scheme to communicate with other intent-connected devices, but that this structure may soon change.
- This is a crucial problem in IoT equipment, such as network antennas, or in cloud systems based on IoT. In order to correctly maintain, install, and operate the core algorithms and dependent passive modules in IoT devices, which consume a lot of energy, if not, we must employ or take into consideration various non-conventional sources of energy as ideal solutions (to power IoT devices or test a cloud when designing cloud systems based on IoT), such as solar power, wind, biomass, and vibration. Researchers will soon need to get active in this field and start working on alternative energy sources. In conclusion, energy is a crucial concern for Internet of Things (IoT) devices (such as smart phones, RFID/embedded chips, and batteries). Future solutions for such gadgets should consider using solar power, wind power, etc., methods to get around this problem.
- Another crucial design consideration that directly impacts quality is fault tolerance. The machine's fault tolerance level will be kept very high to ensure that the device continues to function even when there are technical difficulties in order to create a system that is ideal (with high precision, efficiently). Even in the event of a technical issue, IoT-based clouds need to have quick recovery. Solar or wind energy can function as a superior substitute for batteries as the ideal option (or for devices that use batteries). Keep in mind that prolonged communication with gadgets necessitates higher power usage. Giving certain devices more energy while sparing others can, however, cause a problem with "energy distributions" among IoT devices. However, we can give all devices the same amount of energy because occasionally only certain devices are in use and some devices are in neutral mode (i.e., not working anything).
- Large network lag from the multihop client-to-cloud route lowers the usability and freshness of incoming data for analysis in time-dependent applications. Besides that, IoT data is very volatile for stringent kind of applications, meaning that it is

only relevant for analysis for a very brief period. The problem with recent past data being temporally irrelevant for analysis is volatility.

- The short-range non-IP networks in the IoT ecosystem have sporadic network connectivity issues. When the network tries to resend the lost information, it usually makes mistakes that make the data incorrect. It is known as validity.
- Visualization refers to the difficulty of giving the end-user a concise and understandable presentation of the gathered knowledge. Because of the enormous complexity of the data and the continuous nature of IoT, visualization loop analysis, and many data kinds.
- There is an incompleteness or consistency in the information known as vagueness when multiple sources of IoT data are joined for meaningful analysis, also known as data ambiguity.
- Very recent data must always be accessible for examination in tight real-time analysis. The issue of determining the timeframe of sensor readings that is adequate for investigation and having that data continuously accessible is known as data volatility.

9.4 Challenges and Future Research Directions for AR–VR Embedded Data Analysis

Even though there has been a lot of research and advancement in the field of augmented reality (AR), there are still a number of problems that need to be resolved [16, 17].

- The amount of information in reality is too much for an AR system to handle. As a result, the gear should be easy to transport, lightweight, and quick enough to display graphics. Another restriction on the uses of AR is the battery life that these sophisticated AR gadgets require. In order to supply accurate markers for AR tracking, system hardware like GPS is required. As such, ask them to be sufficiently accurate and dependable. For the use of AR to be practical, several hardware challenges must be overcome. The majority of the time, AR systems collect a large amount of data, which requires software to filter, store, and show in a convenient manner [16].
- All wearable technology needs to be made lighter, smaller, and more user-friendly. Researchers working on AR systems must also take into account other difficulties like response time lags and hardware or software malfunctions.
- Registration mistake is one drawback of AR systems. The study of occlusion detection in AR systems is ongoing. Various tracking techniques are analyzed, and potential paths for tracking research are highlighted. These possibilities enable researchers to successfully use information contained in video frames or creatively combine vision-based techniques with other sensors. Incorporating a recognition system is crucial for obtaining a reference representation of the real environment. Although it is mostly a top-down procedure and challenging to deal with object

dynamics and evaluate many hypotheses, further research in this field may yield encouraging results. Building a widespread middleware to support the AR system is a difficult task [18].

- Some common challenges and research opportunities are found during the implementation of AR systems for learning industries like university. Participants can suggest augmented reality (AR) solutions for important industrial difficulties faced by the industry. However, it is necessary to test actual AR applications in the maritime sector. Recent advancements in wearable augmented reality headsets, like Microsoft HoloLens, which offers educational institutions an affordable way to create AR-based application programming interfaces, will have a big positive impact on the learning industry (APIs). The Idealab is a venue where cyber-physical system research is ongoing. It is feasible to research the efficacy of cyber-physical systems by fusing IoT with industrial augmented reality (IAR) [18].
- AR has a severe impact and lot of future research opportunities in maritime industry. An AR research opportunity to be carried out in partnership with Offshore Simulation Centre will identify how to set up a training facility for personal manufacturing items for large maritime asset makers in maritime industry. Together with Idealab, such a facility will provide short-cycle training experiences that will prepare the staff to carry out the intricate manufacturing process. To prevent any downtime costs, such processing lines need a precise repair and maintenance schedule [17, 18].
- Big data and augmented reality (AR) convergence present both practical and technical concerns. Big data issues include heterogeneity, scale, and complexity obstruct every step of the process, from data collection through aggregation and analysis. Additionally, lengthy calibration, imperfect reference modeling, and inadequate environmental sensing limit the use of AR. Huge-scale data processing typically requires longer periods of time because the data is so large and severely fragmented. When the minimizing tendency in AR devices clashes with the expanding volume and fragmentation of large data, the issue becomes even more difficult. The cloud can conduct computationally complex activities within a set time limit and store a lot of data [19].
- Privacy issues are becoming more and more of a problem in the context of both AR and large data. AR often needs to access users' personal data and record their location in order to deliver a personalized recommendation. Data analysis can be used to uncover hidden links and models when data is combined from several data sources. The issue of privacy is one that involves both technology and society. To prevent improper use of personal data and harmful AR applications, strong laws and restrictions may be necessary. Differential privacy is a technically feasible method of obtaining data with a low privacy risk, but the information is too condensed to be practical in everyday life. Additionally, it is unfit for handling dynamically changing data [19].
- Big data and augmented reality cannot be combined in a simple way to provide intelligent interpretation. However, a cooperative effort to accept augmented reality information and offer native APIs for interpreting semantically labeled

data from all data sources is a potential resolution. A crucial first step is the adoption of a common data format, such as Augmented Reality Markup Language (ARML) [19].

9.5 Challenges and Future Research Directions for Big Biological Data Analysis

Data on biological sequences is expanding exponentially. Big biological data must be accessible for evolutionary investigations to be successful. For the first time, scientists are able to research the driving forces behind how complete genomes evolve. Therefore, we are living in an exciting time for evolutionary biology [20]. Big biological data analysis usually requires long time and facing different issues. Those issues are listed below,

- Error-prone sequencing technologies are used to create biological data. Algorithms will therefore be made extremely complicated in order to deal with these flaws and uncertainty.
- Big biological data analysis problems demand a lot of computation, and the accompanying techniques are polynomially time-consuming [21].
- Due to the inherent computational complexity, several biological data analysis activities require both a lot of data and a lot of computing power. High performance computing (HPC) might offer a useful tool to address these issues [22].
- In a typical sequence alignment facility, hundreds of TB of sequence data are produced each day. Such a vast volume of data presents problems for hardware support as well as computer scientists' ability to analyze data effectively and efficiently.
- For the past few decades, there have been many options for computing platforms that can be used to analyze biology and medical related data. The choices provide numerous approaches to improve productivity or have the ability to apply biological data analysis methods. There are many options, including typical platforms with multicore parallelism, fast computing clusters, cloud platform, and GPUs. Algorithms can also be implemented in a wide range of programming models, in addition. Choices for programming models frequently range widely because of the added emphasis on productivity and performance. In order to take advantage of the variety of parallelism on computer platforms, people have been working on programming biological data analysis applications using common programming models as OpenMP, OpenCL, MPI, and MapReduce (Hadoop). For scientists and engineers working on biological data processing, the variety of architectures and programming methods offers both benefits and difficulties [23].
- There are two key challenges when using architecture-aware optimization techniques with biological data analytics systems. In order to integrate the algorithms within the architecture characteristics, it is typically required to carefully adjust

or even redesign them first. For instance, GPUs are good for enormous data parallelism, but they have major problems with erratic memory access and processing patterns. Unfortunately, many applications for biological data analytics use erratic data structures and the graph representation used by majority genome assembly techniques. It takes a lot of research to examine and refine those algorithms to make them compatible with the GPU architecture.

- Workflow optimization is necessary due to application dependency in addition to speed enhancement and optimizations that consider architecture. Workflow optimization is primarily used to simplify deployment in a special environment and lessen the expenses associated with moving data among different programs. For various common workflows, scientists have been doing their work in this field.
- Applications using large amounts of biological data can be run on a single CPU or typically unable to meet the performance requirement is an accelerator. In order to scale up biological applications, researchers have been using a cluster that has a substantial number of compute machines. Keep in mind that, generally speaking, scalability refers to a computing environment that includes several processors which are loosely connected over the exact chip. They could be separate processors inside a server, GPUs and CPUs. Utilizing task parallelism, this kind of analytics applications are extremely expandable to several nodes. For instance, in short read alignment, each node can often keep the complete index data structure before processing the reads that are given to it. Different nodes are independent of one another. Instead, due to dependency, some biological data analytics applications make task parallelism challenging to implement. The use of exquisite data parallelism should be investigated in order to expand the utilizations to many nodes. Genome assembly is one of the typical applications that falls under this heading.
- Similar to other high performance computing (HPC) applications, biological data analytics likewise has to deal with the difficulty of programming productivity. We can explore the difficulties to programming productivity based on cutting-edge HPC technology. Models of shared-memory parallel programming that are prevalent today contain OpenMP and POSIX Threads (Pthreads). However, given that many core recently, architectures and those programming models have emerged are either poorly supported or inappropriate due to the hardware's distinct qualities. GPUs, for instance, use CUDA or OpenCL for programming. OpenMP and Pthreads can be supported by Xeon Phi but support the use of Intel Cilk Plus by developers. Moreover, Xeon Phi features a set of 512-bit SIMD intrinsic to use video processing units, which is the secret of Xeon Phi's great performance. The benefit of using such programming languages is that they have the ability to benefit from architecture-aware optimizations to make the most of hardware capabilities. Nevertheless, the drawbacks are programming complexity and poor portability. MPI. On supercomputers, scientists use MPI model for biology related data analysis in distributed environment. Meanwhile, new programming models, such MapReduce and Spark, are introduced for large-scale distributed computing for the rigorous needs of expandability and fault tolerance. These distributed programming frameworks facilitate programming and increase scalability [24].

However, based on the above issues, there are a list of future research works can be done in this area.

- Using steady state and time series data stored in the cloud, one can intend to investigate and assess new techniques using frequent subgraph mining (FSM) for genome-wide pattern identification and network inference in the future. Additionally, researchers will evaluate network inference strategies utilizing Apache Spark and Apache Hadoop.
- It is possible to use the features of big data and cognitive computing to better comprehend the issues around information security, privacy, and trust. The first application of cognitive systems employing the advantages of big data analytics in the healthcare sector provides the way for this to be reproduced in other paradigms of scientific and industrial sectors. Although IBM Watson has achieved success in the healthcare industry, it has also branched out into a number of other industries, including advertising, customer engagement, education, financial services, media, and much more, where it is collaborating with top businesses to help them make better decisions. Spread of the advantages provided by significant technological advancements big data analytics and cognitive computing can provide significant challenges. Small and medium-sized businesses (SMEs). SMEs lack the financial resources necessary to employ such technologies, and the technical expertise and skilled labor are equally important. Future research will therefore concentrate on the use of cognitive computing and big data analytics to SMEs. Utilizing cloud services is one such method of computing. Enterprise resource planning (ERP) is formerly solely used by large businesses, however with the advent of cloud-based ERP services are no longer a specialty item. Researchers might look into the possibilities of using cloud computing as a quicker medium for the widespread application of big data analytics and cognitive computing.
- Big biological data analysis techniques also help us to do prediction of the dangerous virus or bacteria diseases (like SARS-CoV-2) through DNA, RNA data analysis, protein–protein interaction data analysis, gene and microarray data analysis, etc. This kind of data analysis includes the creation of a new human-virus or human-bacteria interactions and leads us to the development of antiviral or antibacterial drugs. However, nowadays, cloud platform gives an immense freedom to work with various softwares (such as Cytoscape, Medusa, Genecodis) to speed up the analysis process of such kind of big biological data [25, 26].

Biological data is still challenging to translate into information that can provide insights in a given context, despite technical developments. In light of this, this article presents a viewpoint on the current application of big data in the field of biology. Big data can be made more widely available and used, bridging the knowledge gap between its development and interpretation.

9.6 Challenges and Future Research Directions for Quantum Data Analysis

A quantum computer can process an exponential number of inputs at once and has memory that is exponentially bigger than its physical size. It may be conceivable for computer systems to improve exponentially. If a true quantum computer existed right now, it might have been able to address problems that are extraordinarily challenging for modern computers, like the massive data issues of the present [27]. The major technical challenge in creating a quantum computer may soon be overcome. Information processing using quantum mechanics is made possible by quantum computing. In a conventional computer, data is displayed as lengthy set of bits that either encrypt a zero or a one. However, a quantum computer makes use of quantum bits [28]. The way a qubit differs from a bit is by encoding the zero and the one into two separate quantum states. It can therefore benefit from the superposition and entanglement phenomena. For instance, twenty-one hundred complex values must be saved in a conventional system to accommodate 100 qubits in quantum systems. On the other way, a device with n qubits is capable of superposition of two states. This means that bigger scale quantum computers can address many big data issues much more quickly than classical computers. It can be used to provide a forecast for the weather, the prediction of natural catastrophes like tsunamis, earthquake, for industry analysis, and many more. Therefore, creating a quantum computer and enabling quantum computing to address big data issues is a task for this generation [29]. There are a number of challenges quantum computers facing while doing data analysis,

- Environmental conditions close to absolute zero are a threat to quantum computing.
- The issues of de coherence (isolate system from outside environment), optimization problem (choose the shortest path), and quantum tunneling (chance of electron disappearing on other side) are some of the main challenges addressed by researchers working in quantum computing.
- Handling and processing of large data or image matrices in quantum computing can be done very efficiently and fast. However, in order to store this large information in quantum system, a good number of qubit system is required. At present, physical development of such kind of large number of qubit system is difficult and expensive.
- There are 5Vs (volume, variety, velocity, veracity, and value) challenges while handling large number of data in petabytes. Unlike classical computers, quantum computers require less number of qubits to handle such challenges due to its superposition property and parallelism. An n-bit classical computer can only analyze one state, whereas an n-qubit computer can analyze 2^n parallel states in a single operation. The conventional computer needs to do this process 2^n times in order to arrive at the same solution as the quantum computer. To handle massive data, this great potential can be easily utilized. As a result, quantum computers are able to handle problems at scales that are larger than those that can be handled by classical computers [30].

Quantum computing is a significant topic with open, difficult problems for scientists in the fields of physics, mathematics, computer science and analyst, engineers, too. This section focuses on a few potential areas of future research among all of these ideas. In areas including robotics, power management, Internet security, distributed quality control, complex chemistry, cash flow analysis, and clinical research, research in quantum computing is only getting started. It could take more than ten years for the application of quantum computing in these fields to reach maturity. However, it is anticipated that within the next five to ten years, traffic optimization will surpass the hype cycle and reach its innovation trigger. Inflated expectations have reached their zenith for adiabatic quantum computing, quantum cryptography, and quantum control. They are projected to be built in less than six years inside the realm of quantum computing. Although they have already exceeded inflated predictions, quantum Internet, quantum neural network, hybrid quantum machine learning, and quantum cryptography are all expected to advance quickly over the next five to ten years. The hype cycle currently has high expectations for quantum-based CIBIL score analysis, detection of defaulters, which will flourish in more than ten years. Quantum algorithms and complexity as well as quantum programming languages and systems have both been the subject of extensive research and may continue to be so for the next five to ten years.

The educational sector is constantly looking for fresh chances to expand and thrive. In order to usher in a new era of technology that includes computers capable of presently impossible computations, ultra-secure quantum networking, and exotic new quantum materials, many higher education institutions have started intensive study with quantum computing. Researchers at Harvard University and University of Oxford focusing on the demonstration of the working principles of quantum magnetic resonance computer that helps for the advancement of technology like GPS navigation, worldwide connectivity, and medical advancements. With the aid of a quantum computer they constructed, MIT researchers are looking into topics including quantum complexity and algorithms, quantum information theory, measurement and control, and applications and links. The fundamentals of quantum theory, free-space and fiber-based quantum communications, superconducting quantum computing, ultra-cold atom quantum simulation, and theories and ideas linked to quantum metrology are the major topics of interest. The future workforce will have new employment opportunities thanks to quantum computing, which has extensive implications for higher education. Additionally, this technology will make it possible to accurately estimate both the expansion of the labor market as a whole and the demand for knowledgeable and competent individuals across all industries. Machine learning will soon benefit from the unleashed power of quantum computing. Since, the domain of quantum computing is in its preliminary phases due to the absence of quantum computers and necessary hardware for its implementation, lack of suitable tools and computation environments for carrying out quantum simulation, quantum machine learning presented a hot challenge in the information processing. However, there is a lot of development in this area, and eventually, it might turn into a gold mine for big data analytics, particularly in the healthcare industry. Since the healthcare industry uses a variety of data forms, including text, images, sensor readings, and streaming data. When combined with AI and the precision-based spine care method, the quantum mechanically entangled light enhances the description of the usefulness of imaging in the diagnosis and treatment of spinal diseases. In real sense, quantum sensors may enhance imaging technology by enabling molecular-level visualization, giving clinicians significantly more accurate imagery to aid them in locating the source of their patients' pain, mapping the effectiveness of their treatments, and even identifying high-risk individuals for the onset and progression of disease. Numerous platforms, including those for genomes, metabolomics, transcriptomics, lipidomics, proteomics, interactomics, epigenomics, and more, have emerged in the age of quantum big data that are worth exploring in relation to spinal illnesses. In the end, it may use quantum computing to handle the aforementioned data platforms, allowing for solid findings about illness causes, providing suggestions for treatment possibilities, and supporting more precise treatments for patients with chronic low back pain. We will be able to create high fidelity simulations using quantum computing that will further combine the cellular, molecular, structural, and overall histopathological subtissue elements with different "panomic" domains, improving and optimizing current treatments, and allowing us to find new markers for disease prevention [30, 31]. Furthermore, the field of regenerative biologics for the spine has faced many obstacles and difficulties, and recent developments are still debatable. However, quantum computing may be useful for more accurate patient and dose selection as well as algorithmic modeling of outcomes and the regenerative process from the molecular to structural level. Finally, using a quantum computing method, it may be possible to more fully comprehend and incorporate the connection of a specific regenerated motion segment with the entire spine and the entire body as a functional, interacting unit. Quantum computing may one day become widespread and essential in all facets of patient care due to the expansion of new data platforms and the demand for data-driven science.

Quantum computing plays a vital role for pattern classification of big data through the implementation of quantum neural network, quantum support vector machine, quantum decision tree, etc. An improved linear and nonlinear binary classifier is implemented using a support vector machine on a quantum computer with exponential speedups in the size of the vectors and the quantity of training instances. The fundamental building block of the algorithm is a non-sparse matrix simulation technique to quickly execute a principal component analysis and matrix inversion for training the data kernel matrix.

Similar to quantum computing, which uses photons as its fundamental building blocks, big data can benefit from using quantum computing to solve its heterogeneity or variety problem [30]. This is because big data is being examined electronically. In the coming years, when quantum computer technology is available, quantum computing will be the trendiest issue for solving big data analytics concerns.

9.7 Conclusion

This chapter is mainly focused on the various challenges, issues, and future research directions of data analysis work along with the different innovative technologies such as biological computing, cloud computing, quantum computing, and so on. It gives you a brief overview of various issues and limitations that today's data analysis techniques have while using along with some popular cutting-edge techniques. This chapter also deals with some interesting future research works that the researchers of this cutting-edge technologies' domain can work upon. Interested readers can also receive a brief overview and interesting references of the data analysis techniques and their related computing paradigm discussed in this book.

References

1. Garg T, Khullar S (2020) Big data analytics: applications, challenges & future directions. In: 2020 8th international conference on reliability, Infocom technologies and optimization (trends and future directions) (ICRITO). IEEE, pp 923–928
2. Varlamis I, Apostolakis I, Sifaki-Pistolla D, Dey N, Georgoulias V, Lionis C (2017) Application of data mining techniques and data analysis methods to measure cancer morbidity and mortality data in a regional cancer registry: the case of the island of Crete, Greece. Comput Methods Programs Biomed 145:73–83
3. Chen CP, Zhang CY (2014) Data-intensive applications, challenges, techniques and technologies: a survey on Big Data. Inf Sci 275:314–347
4. Yaqoob I, Chang V, Gani A, Mokhtar S, Hashe, IAT, Ahmed E, Anuar N, Khan SU (2016) WITHDRAWN: information fusion in social big data: foundations, state-of-the-art, applications, challenges, and future research directions
5. Khan S, Shakil KA, Alam M (2017) Big data computing using cloud-based technologies: challenges and future perspectives. Networks of the future, pp 393–414
6. Anagnostopoulos I, Zeadally S, Exposito E (2016) Handling big data: research challenges and future directions. J Supercomput 72(4):1494–1516
7. Nielsen MA, Chuang IL (2000) Quantum computation and quantum information. Cambridge University Press, New York, USA
8. Assuno MD, Calheiros RN, Bianchi S, Netto MAS, Buyya R (2015) Big data computing and clouds: trends and future directions. J Parallel Distrib Comput 79:3–15
9. Hashem IAT, Yaqoob I, Badrul Anuar N, Mokhtar S, Gani A, Ullah Khan S (2014) The rise of big data on cloud computing: review and open research issues. Inf Syst 47:98–115
10. Wang L, Shen J (2013) Bioinspired cost-effective access to big data. In: International symposium for next generation infrastructure, pp 1–7
11. Lan K, Wang DT, Fong S, Liu LS, Wong KK, Dey N (2018) A survey of data mining and deep learning in bioinformatics. J Med Syst 42(8):1–20
12. Pundir Y, Sharma N, Singh Y (2016) Internet of things (IoT): challenges and future directions. Int J Adv Res Comput Commun Eng 5(3):960–964
13. Dey N, Wagh S, Mahalle PN, Pathan MS (eds) (2019) Applied machine learning for smart data analysis. CRC Press
14. Hussein AH (2019) Internet of things (IOT): research challenges and future applications. Int J Adv Comput Sci Appl 10(6)
15. Bansal M, Chana I, Clarke S (2020) A survey on IoT big data: current status, 13 V's challenges, and future directions. ACM Comput Surv (CSUR) 53(6):1–59

16. Vargas DGM, Vijayan KK, Mork OJ (2020) Augmented reality for future research opportunities and challenges in the shipbuilding industry: a literature review. Procedia Manuf 45:497–503
17. Mekni M, Lemieux A (2014) Augmented reality: applications, challenges and future trends. Appl Comput Sci 20:205–214
18. Yu T, Wang X, Shami A (2017) Recursive principal component analysis-based data outlier detection and sensor data aggregation in IoT systems. IEEE Internet Things J 4(6):2207–2216
19. Bermejo C, Huang Z, Braud T, Hui P (2017) When augmented reality meets big data. In: 2017 IEEE 37th international conference on distributed computing systems workshops (ICDCSW). IEEE, pp 169–174
20. Yin Z, Lan H, Tan G, Lu M, Vasilakos AV, Liu W (2017) Computing platforms for big biological data analytics: perspectives and challenges. Comput Struct Biotechnol J 15:403–411
21. O'Driscoll A, Daugelaite J, Sleator RD (2013) 'Big data', Hadoop and cloud computing in genomics. J Biomed Inform 46(5):774–781
22. Schadt EE, Linderman MD, Sorenson J, Lee L, Nolan GP (2011) Cloud and heterogeneous computing solutions exist today for the emerging big data problems in biology. Nat Rev Genet 12(3):224–224
23. Dall'Alba G, Casa PL, Abreu FPD, Notari DL, de Avila e Silva S (2022) A survey of biological data in a big data perspective. Big Data
24. Byron K (2017) Big data analytics in computational biology and bioinformatics. Dissertation, New Jersey Institute of Technology
25. Ow GS, Tang Z, Kuznetsov VA (2016) Big data and computational biology strategy for personalized prognosis. Oncotarget 7(26):40200
26. Soroushmehr SR, Najarian K (2022) Transforming big data into computational models for personalized medicine and health care. Dialog Clin Neurosci
27. Pandey A, Ramesh V (2015) Quantum computing for big data analysis. Indian J Sci 14(43):98–104
28. Outeiral C, Strahm M, Shi J, Morris GM, Benjamin SC, Deane CM (2021) The prospects of quantum computing in computational molecular biology. Wiley Interdiscip Rev: Comput Mol Sci 11(1):e1481
29. Gill SS, Kumar A, Singh H, Singh M, Kaur K, Usman M, Buyya R (2022) Quantum computing: a taxonomy, systematic review and future directions. Softw: Pract Exp 52(1):66–114
30. Shaikh TA, Ali R (2016) Quantum computing in big data analytics: a survey. In: 2016 IEEE international conference on computer and information technology (CIT). IEEE, pp 112–115
31. Mallow GM, Hornung A, Barajas JN, Rudisill SS, An HS, Samartzis D (2022) Quantum computing: the future of big data and artificial intelligence in spine. Spine Surg Relat Res 6(2):93–98

The manufacturer's authorised representative in the EU is Springer Nature Customer Service Centre GmbH, Europaplatz 3, 69115 Heidelberg, Germany. If you have any concerns regarding our products, please contact ProductSafety@springernature.com

Printed and bound by CPI Group (UK) Ltd, Croydon, CR0 4YY
30/09/2026
02225315-0003